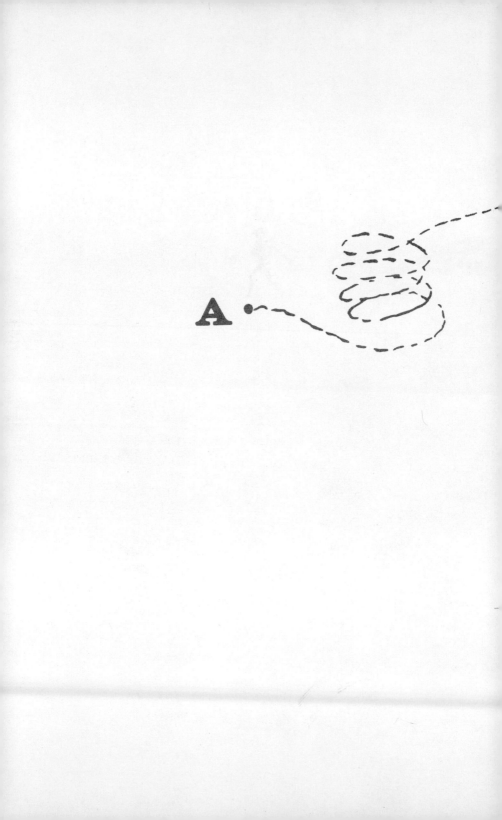

A

3.1416
And All That

by

PHILIP J. DAVIS
Professor of Applied Mathematics
Brown University
Providence, Rhode Island

and

WILLIAM G. CHINN
City College of San Francisco
San Francisco, California

Simon and Schuster
New York

TO OUR FAMILIES

CONTENTS

ACKNOWLEDGMENTS

Most of the articles in this book originally appeared in *Science World*. Grateful acknowledgment is made to *Science World* for permission to use them here. We thank the *Scientific American* for permission to reprint the article entitled "Number."

The authorship of the individual articles is as follows:
Philip J. Davis: 1, 3, 4, 7, 8, 9, 12, 14, 15, 18, 22, 23, 24.
William G. Chinn: 2, 5, 6, 10, 11, 13, 16, 17, 19, 20, 21.

The first author would like to acknowledge his indebtedness to Professor Alexander Ostrowski and to Drs. Barry Bernstein and John Wrench for information incorporated in some of his articles.

The second author would like to acknowledge his indebtedness to Professor G. Baley Price for permission to use materials from one of his articles.

FOREWORD

LYTTON STRACHEY tells the following story. In intervals of relaxation from his art, the painter Degas used to try his hand at writing sonnets. One day, while so engaged, he found that his inspiration had run dry. In desperation he ran to his friend Mallarmé, who was a poet. "My poem won't come out," he said, "and yet I'm full of excellent ideas." "My dear Degas," Mallarmé retorted, "poetry is not written with ideas, it is written with words."

If we seek an application of Mallarmé's words to mathematics we find that we shall want to turn his paradox around. We are led to say that mathematics does not consist of formulas, it consists of ideas. What is platitudinous about this statement is that mathematics, of course, consists of ideas. Who but the most unregenerate formalist, asserting that mathematics is a meaningless game played with symbols, would deny it? What is paradoxical about the statement is that symbols and formulas dominate the mathematical page, and so one is naturally led to equate mathematics with its formulas. Is not Pythagoras' Theorem $a^2 + b^2 = c^2$? Does not the Binomial Theorem say that $(a + b)^2 = a^2 + 2ab + b^2$? What more need be said—indeed, what more can be said? And yet, as every devotee who has tried to be creative in mathematics knows, formulas are not enough; for new formulas can be produced by the yard, while original thought remains as remote as hummingbirds in January.

To appreciate mathematics at its deeper level we must pass from naked formulas to the ideas that lie behind them.

In the present volume, we have selected for reprinting a number of pieces that appeared principally in *Science World,* a periodical with a wide circulation among students and teachers. In writing these articles it was our aim to deal with a number of diverse areas of current mathematical interest and, by concentrating on a limited aspect of each topic, to expose in a modest way the mathematical ideas that underlie it. It has not been possible, in the few pages allotted to each essay, to present the topics in the conventional textbook sense; our goal has been rather to provide a series of appetizers or previews of coming attractions which might catch the reader's imagination and attract him to the thoroughgoing treatments suggested in the bibliographies. Each article is essentially self-contained.

What reason can we put forward for the study of mathematics by the educated man? Every generation has felt obliged to say a word about this. Some of the reasons given for this study are that mathematics makes one think logically, that mathematics is the Queen of the Sciences, that God is a geometer who runs His universe mathematically, that mathematics is useful in surveying fields, building pyramids, launching satellites. Other reasons are that life has become increasingly concerned with the manipulation of symbols, and mathematics is the natural language of symbols; that "Euclid alone has looked upon beauty bare"; that mathematics can be fun. Each of these has its nugget of truth and must not be denied. Each undoubtedly can be made the basis of a course of instruction.

But we should like to suggest a different reason. Mathematics has been cultivated for more than four thousand years. It was studied long before there were Democrats and Republicans or any of our present concerns. It has flourished in many lands, and the genius of many peoples has added to its stock of ideas. Mathematics dreams of an order which does not exist. This is the source of its power; and in this dream it has exhibited a lasting quality that resists the crash of empire and the pettiness of small minds. Mathematical thought is one of the great human achievements. The study of its ideas, its past and its present, can enable the individual to free himself from the tyranny of time and place and circumstance. Is not this what liberal education is about?

Summer, 1968

1

THE PROBLEM THAT
SAVED A MAN'S LIFE

TRUTH IS STRANGER THAN FICTION—even in the woild of mathematics. It is hard to believe that the mathematical ideas of a small religious sect of ancient Greece were able to save—some twenty-five hundred years later—the life of a young man who lived in Germany. The story is related to some very important mathematics and has a number of strange twists.*

The sect was that of the Pythagoreans, whose founder, Pythagoras, lived in the town of Crotona in southern Italy around 500 B.C. The young man was Paul Wolfskehl, who was professor of mathematics in Darmstadt, Germany, in the early 1900's. Between the two men lay a hundred generations, a thousand miles, and many dead civilizations. But between the two men stretched the binding cord of a mathematical idea—a thin cord, perhaps, but a strong and enduring one.

Pythagoras was both a religious prophet and a mathematician. He founded a school of mathematics and a religion based upon the notion of the transmigration of souls. "All is number," said Pythagoras, and each generation of scientists discovers new reasons for

---

* I have the story about the young man—Paul Wolfskehl—from the renowned mathematician Alexander Ostrowski. Professor Ostrowski himself heard the story many years ago and claims there is more to it than mere legend.—P.J.D.

1

thinking this might be true. "Do not eat beans, and do not touch white roosters," said Pythagoras, and succeeding generations wonder whatever was in the man's head. "The square on the hypotenuse is equal to the sum of the squares on the sides," said Pythagoras, and each generation of geometry students has faithfully learned this relation—a relation that has retained its importance over the intervening years.

Paul Wolfskehl was a modern man, somewhat of a romantic. He is as obscure to the world as Pythagoras is famous. Wolfskehl was a good mathematician, but not an extremely original one. The one elegant idea he discovered was recently printed in a textbook without so much as a credit line to its forgotten author. He is principally remembered today as the donor of a prize of 100,000 marks —long since made worthless by inflation—for the solution of the famous unsolved "Fermat's Last Problem." But this prize is an important part of the story, and to begin it properly, we must return to the square on the hypotenuse.

The Theorem of Pythagoras tells us that if a and b are the lengths of the legs of a right-angled triangle and if c is the length of its hypotenuse, then

$$a^2 + b^2 = c^2.$$

The cases when $a = 3$, $b = 4$, $c = 5$ and when $a' = 5$, $b = 12$, $c = 13$ are very familiar to us, and probably were familiar to the mathematicians of antiquity long before Pythagoras formulated his general theorem.

The Greeks sought a formula for generating all the integer (whole number) solutions to Pythagoras' equation, and by the year 250 A.D. they had found it. At that time, the mathematician Diophantus wrote a famous book on arithmetic (today we might prefer to call it the "theory of numbers"), and in his book he indicated the following: If m and n are any two integers, then the three integers $m^2 - n^2$, $2mn$, and $m^2 + n^2$ are the two legs and the hypotenuse of a right triangle. For instance, if we select $m = 2$, $n = 1$, we obtain 3, 4, and 5; if we select $m = 3$, $n = 2$, we obtain 5, 12, 13; and so on.

Our story now skips about fourteen hundred years—while the

world of science slept soundly—to the year 1637, a time when mathematics was experiencing a rebirth.

There was in France a king's counselor by the name of Pierre Fermat. Fermat's hobby was mathematics, but to call him an amateur would be an understatement, for he was as skilled as any mathematician then alive. Fermat happened to own an edition of Diophantus, and it is clear that he must have read it from cover to cover many times, for he discovered and proved many wonderful things about numbers. On the very page where Diophantus talks about the numbers $m^2 - n^2$, $2mn$, and $m^2 + n^2$, Fermat wrote in the margin: "It is impossible to separate a cube into two cubes, a fourth power into two fourth powers, or, generally, any power above the second into two powers of the same degree. I have discovered a truly marvelous demonstration which this margin is too narrow to contain."

Let's see what Fermat meant by this. By Pythagoras' Theorem, a square, c^2, is "separated" into two other squares, a^2 and b^2. If we could do likewise for cubes, we would have $c^3 = a^3 + b^3$. If we could do it for fourth powers, we would have $c^4 = a^4 + b^4$. But Fermat claims that this is impossible. Moreover, if n is any integer greater than 2, Fermat claims, it is impossible to have $c^n = a^n + b^n$, where $a, b,$ and c are positive integers. He left no indication of what his demonstration was.

From 1637 to the present time, the most renowned mathematicians of each century have attempted to supply a proof for Fermat's Last Theorem. Euler, Legendre, Gauss, Abel, Cauchy, Dirichlet, Lamé, Kummer, Frobenius, and hosts of others have tackled it. In the United States, and closer to our own time, L. E. Dickson, who was professor at the University of Chicago, and H. S. Vandiver, a professor at the University of Texas, have gone after the problem with renewed vigor and cunning. But all have failed. The best that has been accomplished to date is to prove that Fermat's Theorem is true for all integers that are greater than 2 and are less than 4,002. This was done in 1954, and required an assist from an electronic computing machine known as SWAC. The figure may be higher today.

Perhaps the most ingenious of all the contributions to the

problem was made by Ernst E. Kummer (1810–1893), who taught mathematics at the University of Berlin. Whereas previous investigators succeeded in proving Fermat's Theorem for isolated values of n (for example, Euler proved it for $n = 3$ and $n = 4$, and Dirichlet proved it for $n = 5$), Kummer was able to prove it for a whole slew of values of n. As a matter of fact, at one point Kummer believed that he had proved the theorem for all values of n, but an unwarranted assumption was later found in his argument.

From Kummer to Wolfskehl is not long in time nor far in distance. As a young student of mathematics, Paul Wolfskehl was attracted to the theory of numbers. The theory is pretty and the methods are difficult. The combination appealed to him. Fermat's Last Theorem was in the air, for the sensation caused by Kummer's near miss had not yet settled down. Wolfskehl tried to prove it. He failed. After all, better mathematicians before him had failed. But it nagged him. He went back to it time and again. He read the works of the masters to see what tools they used and what they were able to accomplish. He made no progress with his own attempts.

In the course of this, he formed a romantic attachment to a young lady. He was disappointed there as well. Now that mathematics and romance were both out the window, he began to feel that life could offer him very little else. He decided to commit suicide. Having made this decision, he went about carrying it out very methodically. He settled his affairs and arranged all important matters. He wrote his will. He fixed upon the method and the very hour of taking his own life. On the last day, he wrote final letters to his friends. Everything was now prepared. A few hours remained till the appointed time. He went into his library, wondering what to do. He took down some mathematical pamphlets from the shelf and fingered them idly.

By pure chance, he opened one of them. It was Kummer's work on Fermat's Last Theorem. As he read the article, he thought he spotted an error in Kummer's work. As a matter of fact, the article begins with a remark that contains a gap in logic. Wolfskehl sat down to check this doubtful point. After all, Kummer was a man with a great reputation, but in the past he had made a very subtle, but crucial, mistake. It was of vital importance to know whether the present argument was correct. One hour passed, two, three hours, while Wolfskehl checked the mathematics. Finally, he was forced to admit that Kummer's argument was completely sound!

When Wolfskehl was through with this job, he reminded himself of his momentous decision to take his own life. But the appointed hour was past. Somehow, he no longer saw the necessity for suicide. From Fermat's Last Theorem had come not only post-

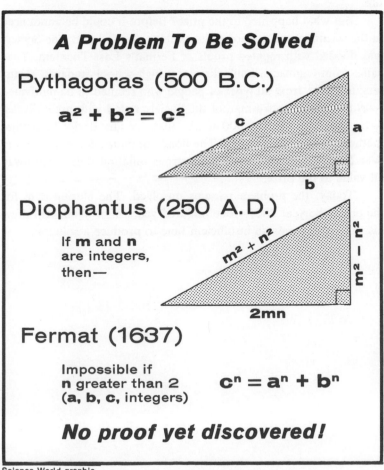

Science World graphic

ponement, but a renewed interest in mathematics and a decision to live. He tore up his last letters and his will.

In 1908, after Wolfskehl died, a new will was found and opened. Here is what it contained: 100,000 marks donated to the Scientific Society of Göttingen—the leading mathematical university of the time—to be awarded to the person who gave a satisfactory proof of the Last Theorem of Fermat. What a noise this caused! There had been mathematical prizes before—1,000 marks, even 5,000 marks. But 100,000 marks! Perhaps this was Wolfskehl's expression of gratitude to the problem that had saved his life.

But what happened to the prize? Before it could be announced in the scientific press, word of its existence leaked out. The Society was flooded with reputed proofs of Fermat's Last Theorem. They came from engineers, bank directors, high-school students, ministers, teachers, from all types of people who seemed to be motivated —as the official spokesman of the Society noted with great disgust —far more by the 100,000 marks than by any desire to further mathematical understanding. The flood continued for years. World War I came, and after that, a disastrous inflation that swept away all value to the prize.

Today, the problem remains unsolved. The closing date for the prize, the year 2007—one hundred years after Wolfskehl drew his will—still seems an insufficient time to produce a solution.

2

THE CODE OF THE PRIMES

SOME 2,400 YEARS AGO, a Greek philosopher by the name of Anaxagoras was foolhardy enough to claim that there were other planetary systems in the universe besides our own. He went further to say that other planetary systems were also inhabited by human beings. At the time he proclaimed these views to the Athenian citizens, the Earth was still thought to be the center of the universe and the sun and moon were considered to be deities. For teaching such revolutionary beliefs, Anaxagoras was condemned to death, although the sentence was later commuted to banishment.

Today, there is a greater tolerance of unconventional ideas in science. Many scientists are convinced that there is a good probability that some forms of life do exist elsewhere in the universe outside our own solar system. But how to detect the presence of these "intelligent beings"? One possibility may lie in the *laser*, a device that generates a new type of ultrapure light. A beam of laser light can be focused so precisely that even after traveling 250,000 miles from earth to moon, it would have spread only to a diameter of two miles.

Could a laser be used to communicate with life on other planetary systems? If so, what kind of information should be transmitted—both to show how intelligent we are and to test the intelligence of our distant friends?

A continuous, regular flashing would not do, because such might be misinterpreted by a foreign observer as merely another natural phenomenon of this planet. An irregular or random flashing would not do, because it would not indicate organized thought. So we would need a set of signals more irregular than a possible natural event and yet regular enough that some systematic design might be observed. One scientist experienced in number theory gave this suggestion: Signal at intervals according to the spacing of the prime numbers!

To see whether this suggestion has merit, we must know something about the prime numbers and how they are spaced in the number system of positive integers (whole numbers) 1, 2, 3, 4, 5, 6, 7 . . . A prime number is an integer that is greater than 1 and is divisible, without a remainder, by no other integer than itself and 1. For example, 2 is a prime number because it cannot be divided by any number other than 2 and 1, and 17 is a prime number because it cannot be divided by any number other than 17 and 1. The number 12, for example, is not a prime, because it can be divided by 1, 2, 3, 4, 6 and 12. Another way of describing primes is to say that they have exactly two *integral factors*. An integral factor is simply a factor (divisor) that is an integer. For example, the number 15, which is not prime, has the factors 1, 3, 5 and 15. All the integers which are greater than 1 and are not prime are called *composite* numbers, since they are "composed" of more than two factors.

Thus, 1 is not considered a prime (number), since $1 = 1 \times 1$, and 1 is the only integral factor of 1; 2 is a prime, because 2 has exactly two integral factors, namely, 1 and 2; 3 is a prime, because 3 has exactly two integral factors, namely 1 and 3; 4 is a composite number, because 4 has three integral factors, namely, 1, 2, and 4; et cetera.

To determine whether a number n is prime or composite we must try all possible divisions with integers and see whether or not the remainder is zero; if we get a remainder 0 only when we divide by 1 or by n, then n is a prime number; as soon as we get a remainder 0 on any other division, we know the number is composite. This task can be reduced by eliminating division by 1 and division by n since we already know that in each case we will get 0 as a remainder. We may further eliminate division by any integer greater than n, since we will always get a remainder n on such division. So we are now faced with, at most $n - 2$ divisions. For small numbers, this is an easy task; we have a small number of divisions to carry out. But when the number is terribly large this process may not be practical and we may want to reduce the task still further. For the next reduction, let's examine how a number may be expressed as a product.

Let $n = ab$. If $a > \sqrt{n}$,

then since $a = \dfrac{n}{b}$,

we have $a = \dfrac{n}{b} > \sqrt{n}$.

From which, $b < \sqrt{n}$.

Now, what does this last inequality tell us? It says that if n has factors, one of which is greater than the square root of n, then the other must be less than the square root of n. All this agrees with our intuitive feeling; it is merely stated clearly in algebraic form. What it does imply too, is that we don't have to try division by integers greater than the square root of n, because if there is a factor greater than the square root of n we would have found it paired with a factor that is less than the square root of n. This is a tremendous reduction in labor; for example, in testing whether 997 is prime, instead of 995 divisions, we can now get by with 31.

This is not all the reduction possible. Notice that a multiple of a number cannot divide into n exactly unless the number itself divides into n exactly. For example, since 6 is a multiple of 3, 6 is not a factor of any number unless the prime 3 is also a factor of the number. This means that we need only try all divisions *by prime numbers* smaller than the square root of n. For $n = 997$, the number of prime numbers less than the square root of n is 11—a substantial reduction from 995. Still, the job is not always easy and many theorems have been developed for the purpose of. determining whether or not a number is prime.

One of these theorems was announced, unproved, in 1770 by Edward Waring and attributed to a John Wilson. The theorem states that a number p is prime if and only if the next integer after the number $(p - 1)(p - 2)$ . . . $2 \cdot 1$ contains p as a factor. Within a few years after its announcement, it was proved by Euler and Lagrange and later a very elegant proof was given by Gauss. The "Wilson Criterion," though true, is not always practical, because the calculation of $(p - 1)(p - 2)$ . . . $2 \cdot 1$ may be quite unwieldy. Another theorem, announced by Joseph Bertrand and proved by Pafnuti Chebyshev, states that between any number and its double, there is at least one prime number. This does not give us a practical means of locating the prime numbers, but the value of such a theorem is that it often leads to other important theorems which may have a direct bearing on the original problem.

Even with a big catalog of theorems we do not have all the answers. Much effort has gone into studying the distribution, or spacing, of primes in the positive integers. If we could unlock the key to this mystery we could predict exactly where the next prime would appear in the sequence of natural numbers and our problem would be solved. A glance at the distribution of the first 100 prime numbers in the table below will indicate the kind of job in store for us.

THE FIRST HUNDRED PRIME NUMBERS														
2	3	5	7	11	13	17	19	23	29	31	37	41	43	47
53	59	61	67	71	73	79	83	89	97	101	103	107	109	113
127	131	137	139	149	151	157	163	167	173	179	181	191	193	197
199	211	223	227	229	233	239	241	251	257	263	269	271	277	281
283	293	307	311	313	317	331	337	347	349	353	359	367	373	379
383	389	397	401	409	419	421	431	433	439	443	449	457	461	463
467	479	487	491	499	503	509	521	523	541					

Irregular spacing of prime numbers poses many problems.

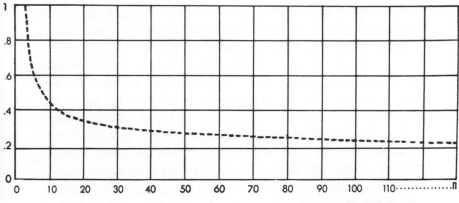

Science World graphic

Number of primes scattered through the number system diminishes gradually among the higher numbers. The curve shows, as an average fraction, rate at which the prime numbers appear.

Apparently there is no observable pattern to help us predict where the next prime number might pop up. The gap between two neighboring prime numbers may widen here and there; then we may find two or three primes packed together, as shown in the table. This erratic behavior in the distribution of primes provides a sufficiently irregular code for interplanetary communication with laser signals. The big point in its favor is that the primality of a number is independent of the system of counting that Planet X may happen to use. In this sense, it is an interplanetary language. All this is based on the idea that any intelligent being would necessarily have developed some number sense—more specifically, that his intellectual curiosity would have forced him to investigate this aspect of number theory. Whether this is a valid assumption remains to be seen. However, since we have dwelt on the erratic behavior of the prime numbers, we should also mention some of their predictable characteristics.

Euclid has already given us the answer as to how many prime numbers there are: namely, an infinite number of them. He did this by first assuming that there is a finite number of prime numbers. Using all of these, he produced a new prime number, thus contradicting his assumption. He concluded that there must be an infinite number of primes. For example, if $a, b, c, \ldots z$ is a complete list of "all" the prime numbers, then adding 1 to the product of all these gives us $(a\ b\ c\ \ldots z) + 1$. This is a new number which does not leave 0 as remainder on division by any of the prime numbers, $a, b, c, \ldots z$ (division by any of the numbers $a, b, c, \ldots z$ will give 1 as remainder). Behind this proof lies a fundamental theorem, stating that if a and b are prime numbers then the product $a\ b$ cannot be divided by any other prime number. A word of caution should be noted: if a, b, c are prime numbers, this does not imply that $(a\ b\ c) + 1$ is a prime number. For example, $(3 \cdot 5 \cdot 13) + 1 = 196$, and 196 is divisible by 2, 4, 7, 14, 49, $\ldots$ All we can say is that $(a\ b\ c) + 1$ is *not* divisible by $a, b,$ or $c;$ however, if the composition involves "all" the prime numbers as Euclid had assumed, then the new number is not divisible by any number in the so-called "complete" list of primes, hence it is not divisible by any composite number made up of these primes, and is therefore a prime number. Since it is shown that we can always produce at least one more prime number no matter how many we already have, the number of primes is infinite.

While we are guaranteed an inexhaustible supply of primes we can also find intervals barren of prime numbers. These may be called *prime deserts*. For example, between 7 and 11, there is a small desert of three numbers which are not prime; between 89 and 97, there is a larger desert of seven composite numbers. The manufacturing of deserts calls for numbers of the form $n(n - 1) \cdot (n - 2) \ldots 2 \cdot 1$ which we met before in connection with Wilson's Criterion. These are the *factorial* numbers, and $n(n - 1) \cdot (n - 2) \ldots 2 \cdot 1$ is abbreviated by the symbol $n!$, so that, for example, 6! means $6 \cdot 5 \cdot 4 \cdot 3 \cdot 2 \cdot 1$, or 720.

Now look at $n = 6! = 720$ and the sequence of integers greater than $n + 1$. We know by the manner in which 6! is composed that 6! is divisible by 2, 3, 4, 5, 6, and that

n divisible by 2 implies $n + 2$ divisible by 2,
n divisible by 3 implies $n + 3$ divisible by 3,

> n divisible by 4 implies $n + 4$ divisible by 4,
> n divisible by 5 implies $n + 5$ divisible by 5,
> n divisible by 6 implies $n + 6$ divisible by 6.

Since $n + 2$ to $n + 6$ are the numbers from 722 to 726 inclusive, we have located a desert containing a minimum of 5 composites. In a similar manner, the interval from $(1,000,001)! + 2$ to $(1,000,001)! + 1,000,001$ will yield a desert of *at least* a million composite numbers, one after another. In general, there is a prime desert of at least n composites in the interval from $(n + 1)! + 2$ to $(n + 1)! + (n + 1)$. So if you are looking for a desert barren of any number of prime numbers, you can go off in a corner and create your own desert—in spite of the guarantee that the supply of primes is inexhaustible.

3

POMPEIU'S MAGIC SEVEN

IT WOULD BE a mistake to think that professional mathematicians spend all their time thinking about deep and difficult problems. The differential equations of magnetohydrodynamics or the properties of infinite dimensional spaces are all very well in their place, but when a mathematician wants to relax, he finds nothing finer than a simple puzzle.

Recently, while browsing in the collected works of the Rumanian mathematician Dmitri Pompeiu, I came across an interesting example of the kind of puzzle that intrigues mathematicians. In 1951, after a lifetime of teaching at the University of Bucharest and just three years before he died, Pompeiu posed the following puzzle in arithmetic and showed how to solve it: ABC,398,246 is a number in the hundreds of millions whose first three digits, A, B and C, are unknown. We are informed that this large number is exactly divisible by 317. The puzzle is to find out what the digits A, B, and C are.

Your first reaction to this puzzle may be one of complete bewilderment. In division, we normally work first with the digits of the dividend that are to the extreme left, but in the puzzle, these are just the ones that are unknown.

Where do we start, to solve such a problem? The secret of success, says Pompeiu, depends on two approaches to the problem: The first is to change the division into a multiplication—in this way

we deal first with the digits to the far right. The second is to use the
following curious fact about the number 7, which is the units digit
of the number 317.

Suppose that 7 is multiplied successively by each of the digits
from 0 to 9, inclusive:

	7	7	7	7	7	7	7	7	7	7
X	0	1	2	3	4	5	6	7	8	9
	0	7	14	21	28	35	42	49	56	63

In these products, forget the tens digits and take a look at the
units digits. They are, respectively, 0,7,4,1,8,5,2,9,6,3. We see that
they are all the digits from 0 to 9, but switched around into a different sequence. In other words, this sequence is a *permutation* of
the digits. This permutation sets up a *one-to-one correspondence*
between the multiplier digits and the products digits, which can be
shown in the following way.

0	1	2	3	4	5	6	7	8	9
↕	↕	↕	↕	↕	↕	↕	↕	↕	↕
0	7	4	1	8	5	2	9	6	3

Not all digits have this remarkable property. For instance, if
6 is used as a multiplicand:

	6	6	6	6	6	6	6	6	6	6
X	0	1	2	3	4	5	6	7	8	9
	0	6	12	18	24	30	36	42	48	54

Ignoring the tens digits in the products, we obtain the correspondence

0	1	2	3	4	5	6	7	8	9
↓	↓	↓	↓	↓	↓	↓	↓	↓	↓
0	6	2	8	4	0	6	2	8	4

In this case, the second line is not a permutation of the first
line, because the odd numbers are not present.

With this in mind, let's return to our puzzle. Now, 317 times some unknown number—call it . . . *RSTUVWXYZ* (we will soon see how many digits are required)—equals *ABC*,398,246. Write this out as a multiplication.

$$
\begin{array}{r}
3\,1\,7 \\
\ldots\;RSTUVWXYZ \\
\hline
\end{array}
$$

$$
\begin{array}{r}
\cdot\;\;\cdot\;\;\cdot \\
\hline
ABC398246
\end{array}
$$

If, in your mind's eye, you think of the multiplication carried out, you will see that 7 times *Z* must be a number that ends in 6. Consulting our one-to-one correspondence, we find that the number *Z* must be 8—there is no other possibility. Writing 8 for *Z,* we can complete a part of the multiplication:

$$
\begin{array}{r}
3\,1\,7 \\
\ldots\;RSTUVWXY8 \\
\hline
2\,5\,3\,6 \\
\ldots\;HJKL \\
\cdot\;\;\cdot\;\;\cdot \\
\hline
ABC398246
\end{array}
$$

At this point, we bring in a second row of unknown digits, written as . . . *HJKL*. What can we say about them? It is clear that $L + 3 = 4$. Therefore, *L* must be 1. Arguing further, 7 times *Y* must be a number that ends in 1. Consulting our correspondence, we find that *Y* must be 3. Again, there is no other choice. After we write 3 for *Y,* we can complete the second line of the multiplication:

$$
\begin{array}{r}
3\,1\,7 \\
\ldots\;RSTUVWX3\,8 \\
\hline
2\,5\,3\,6 \\
9\,5\,1 \\
\ldots\;DEFG \\
\cdot\;\;\cdot\;\;\cdot \\
\hline
ABC398246
\end{array}
$$

Bringing in a third row of unknown digits, . . . *DEFG,* we notice that $5 + 5 + G$ must be a number that ends in 2. It follows that G must be 2 and X must be 6.

This should indicate the path toward the complete solution of the puzzle. Work the puzzle through and see if you come out with this answer: $A = 3, B = 1, C = 3$.

Pompeiu's solution depends crucially upon the permuting power of the multiplication by 7. But is 7 the only digit that has this power? A little experimenting with numbers will show that it is not. Two others are easily found: 3 and 9. Multiplication by 3 brings about the correspondence

0	1	2	3	4	5	6	7	8	9
↕	↕	↕	↕	↕	↕	↕	↕	↕	↕
0	3	6	9	2	5	8	1	4	7

while multiplication by 9 brings about the correspondence

0	1	2	3	4	5	6	7	8	9
↕	↕	↕	↕	↕	↕	↕	↕	↕	↕
0	9	8	7	6	5	4	3	2	1

Notice how this permutation just switches around the positive integers.

There is one multiplier that is so simple we have overlooked it. It is the number 1. Multiplication by 1 brings about the permutation

0	1	2	3	4	5	6	7	8	9
↕	↕	↕	↕	↕	↕	↕	↕	↕	↕
0	1	2	3	4	5	6	7	8	9

You may not think of this as a permutation at all. But it is a permutation in the sense that the lower line contains all the digits. It is frequently called the *identity* permutation. It doesn't change a thing.

We now suspect that in addition to 7, we could make and

solve a Pompeiu puzzle with a multiplier that ends in a 1, 3 or 9. This is indeed the case. Try this one: the number $A,BC8,682$ is exactly divisible by 213. Find the digits, A, B and C.

What lies behind the permuting power? What is strange about the numbers 1,3,7 and 9? It seems to be such an odd assortment. Here is the secret: 1,3,7 and 9 are precisely those digits that are *relatively prime* to 10. This means that the numbers 1,3,7, and 9 have no factors (other than 1) in common with 10. How does the number 10 get into the act? It gets in because we ignored the tens digits in the products when we set up our permutations. Another way of phrasing it is that we subtracted away from the products all multiples of ten that were less than the product in question.

The process of subtracting away multiples of a fixed number is called *residue arithmetic*. Here is an example of how it works: Suppose our fixed number is 9 (and not 10 as before). Then subtracting away multiples of 9 from, say, 11, 18, and 58 would leave 2, 0, and 4 respectively, since $11 = (9 \times 1) + 2$, $18 = (9 \times 2) + 0$, $58 = (9 \times 6) + 4$.

The magic property of 1, 3, 7 and 9 is but a special instance of a more general phenomenon: *If we agree to subtract multiples of a fixed number N, then multiplication of the digits by any fixed number relatively prime to N will permute the digits.* This theorem cannot be demonstrated here, but its proof can be found in Birkhoff and MacLane, *A Survey of Modern Algebra,* and Hardy and and Wright, *The Theory of Numbers* (see Bibliography).

Here is one example: Suppose we agree to perform residue arithmetic with respect to 9. Select 5 (a number relatively prime to 9) as a multiplicand.

	5	5	5	5	5	5	5	5	5
X	0	1	2	3	4	5	6	7	8
	0	5	10	15	20	25	30	35	40

Subtracting out multiples of 9:

0	5	1	6	2	7	3	8	4

This leads to the permutation

0	1	2	3	4	5	6	7	8
↕	↕	↕	↕	↕	↕	↕	↕	↕
0	5	1	6	2	7	3	8	4

If we had selected a number N which is itself prime, all digits are automatically relatively prime to it, and hence all multiplicands lead to a permutation.

The study of residue arithmetic leads to many remarkable results. On the one hand, there are numerous tricks and puzzles. On the other, there are some of the most profound theorems in the theory of numbers. You will meet *finite fields*—systems of numbers in which addition, subtraction, multiplication, and division are possible, but in which there are no fractions! If you go far enough, you will meet *finite geometries*. These are geometrical systems which have only a finite number of points and a finite number of lines and, strangely enough, find application in the statistical theory of planning experiments!

4

WHAT IS AN ABSTRACTION?

MATHEMATICIANS play many roles. They may be puzzle solvers, computers, logicians, theorem provers. They are also abstractors, and this is perhaps fundamental to their other roles. The famous English physicist Paul Dirac once said, "Mathematics is the tool specially suited for dealing with abstract concepts of any kind, and there is no limit to its power in this field." Now, just what does the word *abstract* mean? The dictionary tells us:

> **abstract** *adj.* **1.** Considered apart from any application to a particular object. **2.** Ideal. **3.** In art. Characterized by non-representational designs. **4.** In math. Used without reference to a thing or things, as the abstract number 2.

This is a beginning, but only that, for dictionaries can't do the whole job. No dictionary definition of a camel ever expressed the complete camelness of that animal. In order really to know what a camel is, you have to see one and feel one and smell one and live around one for a few years. As much is true of the word *abstract*. We must live with the word a while in order to get its flavor.

The word *abstract* is perhaps most familiar to us these days through abstract art. Abstract art is frequently geometrical in character. There may be no recognizable figures such as apples or trees.

The artist seems to be concerned with the bones of things rather than with the flesh and blood of things. Abstract art has been around for more than fifty years, but in a deeper sense it is much more than fifty years old, for all art is abstract. The portrait of a man is not the same as a man. It is only a statement of certain aspects of a man in which the artist is interested. The artist *abstracts* (or extracts) these characteristics from his total experience of the man.

There is a parallel situation in mathematics. A special brand of mathematics has been developed in the last half century known as *abstract mathematics*. It deals with such things as abstract sets and abstract algebra. But in reality, all mathematics is abstract. When the first arithmeticians in the remote Stone Age looked at two apples and two trees and two men, and extracted from these pairs of things the abstract notion of the number two, mathematics was born. Abstraction operates like the dehydrator that converts a vat full of soup into a package of soup powder. It results in a very powerful stuff.

Mathematics feeds upon abstractions. Numbers of all sorts are abstractions; straight lines, circles, geometric figures of all kinds are abstractions. The notion of dimension is an abstraction. The arithmetic operations such as addition and subtraction are abstractions. And there are abstractions of abstractions, such as sets of numbers or families of curves. There are even abstractions of abstractions of abstractions.

To carry out his role of abstractor, the mathematician must continually pose such questions as "What is the common aspect of diverse situations?" or "What is the heart of the matter?" He must always ask himself "What makes such and such a process tick?" Once he has discovered the answer to these questions and has extracted the crucial simple parts, he can examine these parts in isolation. He blinds himself, temporarily, to the whole picture, which may be confusing.

In this article we will show, by discussing a simple example, how the process of abstraction is carried out. The example is a modern one and has the advantage that it is not part of one's second nature, the way numbers are.

Take a look at Figures 1A and 1B, and ask yourself what these figures have in common. At first glance it may appear that they have nothing in common. Figure 1A seems to be a series of boxes within boxes, while Figure 1B might be a sketch of a pearl necklace. Figure 1A seems complicated, Figure 1B somewhat simpler in its organization.

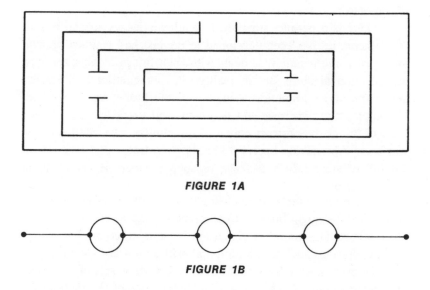

FIGURE 1A

FIGURE 1B

Yet, there is one respect in which these figures are completely identical. It may have occurred to you that Figure 1A resembles a maze or a labyrinth. The object of a maze is to start from the outside, find one's way to the innermost chamber, and then get out. We walk down the corridors, more or less at random, trying to find an entrance that will take us in one more layer. If we have made a complete investigation of the maze, we are in a position to describe it verbally. The magician, giving his instructions to Aladdin at the mouth of the cave, might have said: "Enter by a door at *A*. You will then find two passageways, one of gold and one of silver (*B* and *C*). Take either passageway, and you will come to a second open door. The second door (*D*) opens onto two halls. The first (*E*) is lined with diamonds, and the second (*F*) is lined with rubies. You may go down either hall, and you will again come to a door (*G*). This leads to two more passageways (*H* and *I*), one of pearl and one of opal, either of which will take you to a door (*J*). This door will admit you to the treasure chamber (*K*) where the lamp and other loot is. To get out, reverse the procedure." (*Figure 2A.*)

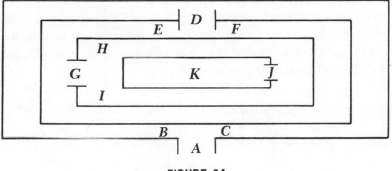

FIGURE 2A

Now suppose we mark Figure 1B as we have marked Figure 2B and regard it too as the diagram of a maze.

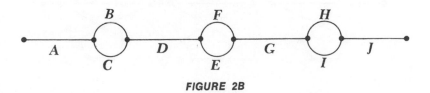

FIGURE 2B

Then, it is easy to see that as far as the varieties of paths that are possible in it, Figures 2B and 2A have the same verbal description. In this respect they are identical. If Figure 2A (or 1A) is regarded as an exact floor plan of the maze, then Figure 1B (or 2B)—which is a lot simpler conceptually—does not carry with it as much information. But in the darkened passageways Aladdin probably was not interested in knowing too much. He did not care how long the passages were or how wide they were, whether they twisted and turned, or whether they wrapped around themselves or crossed over themselves like a cloverleaf in a superhighway. All he was interested in was the way the respective passages played into one another and led to the treasure chamber. If he was interested in this aspect and this aspect only, then he could have done with Figure 1B, despite the fact that it is not the exact geometrical counterpart of the maze.

A mathematical object that interests us only in the possible traversings that it represents is known as an *abstract graph*. (The word *graph* is more commonly used in mathematics to describe data that have been plotted. This is not what is meant here.) Figure 3 gives additional examples of graphs.

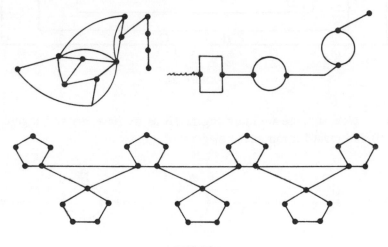

FIGURE 3

This type of abstraction is really quite commonplace. For instance, it occurs on transportation maps. In New York City, a map of the subway system is posted in every subway car. The relative distances are not accurate, the depth of the tunnel is not indicated, the locations of entrances and exits to the subway have been deleted. But for indicating what train goes where, the order of the stations, and where to transfer, the map is complete.

The process of abstraction can be continued further and away from geometry completely. In Figure 1B we designated the places where alternative passages occur by small dots. Suppose that instead of labeling the passageways with letters as we did in Figure 2B, we label the dots instead:

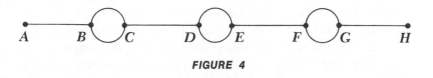

FIGURE 4

The lines that connect the dots are only symbolic. We could really dispense with them. It would serve our purpose as well to write

    AB, 1 *BC,* 2 *CD,* 1 *DE,* 2 *EF,* 1 *FG,* 2 *GH,* 1

indicating that one passageway connects *A* and *B,* two separate passageways connect *B* with *C,* et cetera. One can even put this information into a table in this way:

	A	*B*	*C*	*D*	*E*	*F*	*G*	*H*
A	0	1	0	0	0	0	0	0
B	1	0	2	0	0	0	0	0
C	0	2	0	1	0	0	0	0
D	0	0	1	0	2	0	0	0
E	0	0	0	2	0	1	0	0
F	0	0	0	0	1	0	2	0
G	0	0	0	0	0	2	0	1
H	0	0	0	0	0	0	1	0

FIGURE 5

Along the side and top of the table, we have indicated all the points in the graph, while in the body of the table we have inserted the number of separate passageways that connect the respective points. The table in Figure 5 is called the *incidence matrix* of the graph. Such tables were brought to prominence by the great French mathematician Henri Poincaré (1854–1912). An incidence matrix is able to describe a graph in a completely arithmetic way.

Here, then, in Figures 1A, 1B and 5, are three different concrete representations of one and the same "traversing" situation. Which representation is the most convenient one to use may depend upon the situation at hand. If, therefore, we are interested in setting up a theory of graphs which describes only the properties of traversings and nothing else, we must proceed in a purely abstract way that takes no notice of any other properties these representations may possess. We start with purely abstract definitions. We assume that we are dealing with a set of things, $A, B, \ldots$, which represent the points at which alternative paths branch off. These points—or *vertices*, as they are called in graph theory—are connected in a certain way by means of "lines." But we are thinking necessarily of geometric lines. The connection of A with B is simply a certain relationship which is possessed by A and B simultaneously.

With this as our raw material, we can proceed to make deductions. Although the original inspiration may have been geometric, we have shed all geometric garments. We have extracted all we need to describe traversings. The resultant theory can be carried out by the laws of the combination of a few symbols.

5

POSTULATES—THE BYLAWS OF MATHEMATICS

HAVE YOU EVER WATCHED ants go about their business? You probably were fascinated by the way each ant goes tunneling through the sand, toting food here and there. I can remember many days spent in this ant-watching occupation. It seems that we could always find ants parading up and down the sidewalks. If not, a picnic would be enough to bring on the insects.

Part of the fascination ants hold for us lies in the orderly way in which each ant goes about its business within an organized community. Because they conduct their affairs in an organized manner, ants are considered "social" insects. Human beings too find themselves thrown together in a variety of organizations or clubs. If the club is small enough we don't need much machinery to run it; but when the club has a sizable membership, as the American Mathematical Society, for example, it needs to be highly organized.

Last year, the American Mathematical Society listed 12,443 members on its roster. To operate efficiently with so many members, the Society governs itself with a set of bylaws. To carry out its many functions, it parcels out the work among committees as required. Even for a relatively small organization, such as a classroom or perhaps a student body, bylaws and committees may be necessary.

To illustrate the care that should go into the consideration of bylaws, let me cite "The Case of the Ants in a Pickle." One day a group of ants met for the purpose of forming a club. The group took up, as the first order of business in this organizational meeting, the questions relating to the ideals and working procedure for the membership. The acting secretary jotted down the following points, which everyone present had agreed were desirable and should be covered by regulations for the members:

> work to be parcelled out; promote cooperative spirit among the members; keep the club from being dominated by a small group; provide for interchange of ideas within working groups; keep working groups from being too unwieldy by limiting their size; distinguish the operations of working groups from operations of the entire membership; provide for communication of ideas between working groups.

To carry out these ideas, the following bylaws were adopted:

1. There will be at least one committee in the club.
2. For any two members of the club, there will be a committee on which the pair will serve together.
3. For any two members, there will be at most one committee on which the same pair will serve together.
4. Each committee will consist of at least three members.
5. No committee will consist of more than three members.
6. Not all members are to serve on the same committee.
7. On any two committees, there will be at least one member serving on both committees.

When you compare these bylaws with each of the points the secretary had listed, you will agree that they conform to the wishes of the group. Unfortunately, in no time at all it was discovered that due to these requirements the club must immediately expel two members because the existing membership was too large. How to get out of this pickle or whom to expel is the question the club must face. As for us, we can ask: How many were present at the organizational meeting? How do we arrive at this number?

To answer these questions, we shall adopt a graphic presentation. Each member of the club can be represented by a point; members belonging to the same committee are indicated by connecting lines between the points. The diagrams in Figure 1 sketch this line of reasoning.

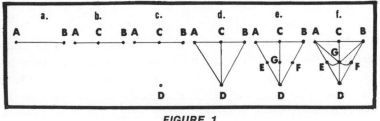

FIGURE 1

In the step-by-step explanation below, the numbers in parentheses refer to the relevant bylaws.

a. Draw *AB;*   (2) between two points is a connection (path), (1) there is at least one connecting path.

b. Locate *C* on *AB;*   (4) any path connects at least three points, (5) no path connects more than three points.

c. Locate *D* outside *AB;*   (6) there is a point outside the path.

d. Draw *DA, DC, DB;*   (2) between two points is a connection.

e. Locate *E* on *DA, F* on *DB, G* on *DC;*   (4, 5) each connection connects exactly three points.

f. Draw *EGB* and *FGA;*   (2, 3, 4, 5) connect each new point to all other points; route the path so that there will be three points on each path.

So far so good. However, *CE, EF* and *FC* are still unconnected. The connection can be made by a circuit tying together *C, E* and *F*. We arrive thus at the final picture as shown in Figure 2. This final picture tells us that in order to meet all the requirements, the club must have *at least* 7 members, *A, B, C, D, E, F, G,* and 7 committees, *ACB, BFD, DEA, AGF, BGE, DGC* and *CFE.*

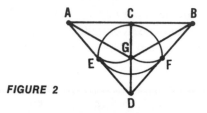

FIGURE 2

Now let's see what happens on the addition of an eighth member, *H,* to the club. By the second bylaw, *A* and *H* must serve together on a committee. By the last bylaw, this committee must have a member in common with *BFD;* this cannot happen, since Bylaw 3 forbids two members to serve together on more than one committee, and *A* has already served once with each of *B, F* and *D.* As we have just shown that this club must have exactly 7 members and as we were told that it was necessary to expel 2 in order to fulfill the requirements, there were originally 9 ants at the meeting.

The example just given illustrates a topic in mathematics known as *finite geometry* (also called *miniature geometry*). Here, the mathematical system consists of a finite number of points, lines, and so on. It may apply to the above examples, or to telephone trunk lines, utility lines, or to electrical circuits. Essentially, the facts of the problem are translated in terms of points, lines, and space as we have done, and the limiting requirements become the postulates of a mathematical system. Usually there is no attempt to interpret what is meant by a *point* or a *line;* they are simply convenient names for some objects, and the mathematician plays a game by the rules set forth as postulates. This is no different from playing baseball by league rules or boxing by the Queensberry rules. In each case, the rules spell out what can or cannot be done without invoking some penalty; the penalty for the mathematician is a fault in logic.

To catch a glimpse of this mathematical game, we can translate the requirements of the bylaws as before, using points to represent club members and lines to represent committees. Notice that in this finite system, we don't worry about the length or straightness of lines, distance between two points, et cetera. We are concerned chiefly about whether certain points lie on certain lines or certain lines go through certain points, so that the symbol AB simply means a line containing the points A and B. Whether a point lies on a certain line or whether a line contains a certain point is merely a question as to whether a certain object comes in contact with another, and these are described as *incidence relations* in geometry. The bylaws are thus translated into these postulates:

Postulate 1. There exists at least one line.

Postulate 2. Any two points lie on a line.

Postulate 3. Any two points lie at most on one line.

Postulate 4. Each line contains at least three points.

Postulate 5. No line contains more than three points.

Postulate 6. Not all points are on the same line.

Postulate 7. Any two lines intersect in at least one point.

From various combinations of these postulates, we arrive at some inescapable consequences—theorems of this particular geometry. For example, from Postulates 2 and 3 we get

I. Any two points lie on one and only one line.

As another example, from Postulates 4 and 5 we get

II. Each line contains exactly three points.

Postulate 7 guarantees at least one point of intersection for any two lines, and Postulate 3 implies that two or more intersections are impossible since each pair of points can lie on at most one line; hence,

III. Any two lines intersect in one and only one point.

Postulate 6 implies that for each line in this set there exists at least one point of the set not lying on the line. Consequently, this outside point together with two points which lie on the line guarantee that

IV. There exist three points not all on the same line.

Notice that Theorems I, II and IV correspond to steps *a, b* and *c* in the diagram.

We may want to look into the behavior of one system of postulates and reinterpret the results in terms of an equivalent system. As a hint of an application, consider all the straight lines in three-dimensional space, or "3-space." To visualize them, think of a coordinate plane on one wall of the room and a coordinate plane on the opposite wall. Two numbers are required to locate a point on each of the planes, so each plane is two-dimensional in the totality of its points. As any point on one plane and any point on the opposite plane determine a line in space, a line in space requires four numbers to fix its location. Hence 3-space is four-dimensional in lines. By studying line geometry in 3-space, then, we arrive at corresponding theorems for points in four-dimensional space; this puts much less strain on our imagination than to try to visualize four dimensions.

Historically, Euclid had done such a beautiful job in assembling the knowledge of geometry available about 300 B.C. that for a long time people considered this body of work almost sacred. It was not until the last century that János Bolyai (1802–1860), Nikolai Lobachevsky (1793–1856), Georg Riemann (1826–1866) and others considered the consequence of changing one of Euclid's postulates—much the spirit of our mathematical game—and in so doing founded *non-Euclidean geometry*. In casting aside the postulate that parallel lines never meet, which was intuitively evident, the immediate reaction may be either that we can no longer find a physical model or that the model does not fit our own physical universe. While this does not normally bother mathematicians, Riemann's geometry—where parallel lines *do* meet—later provided Albert Einstein with the tool for his curvilinear universe and his theory of relativity. In this atomic age we are, of course, caught squarely in the repercussion of Einstein's discoveries, which fundamentally echo the results of a postulate game.

6

THE LOGICAL LIE DETECTOR

SIR ARTHUR CONAN DOYLE was a master at weaving tangled tales of intrigue, whose plots were eventually unraveled by the cold logic of Sherlock Holmes. "Elementary, my dear Watson," says Sherlock, and so it is, if you can keep track of the numerous clues leading to the one unescapable conclusion. If you feel up to tackling this type of mystery, join some mathematical detectives and take a crack at the task of tracking down a murderer. The story goes as follows:

Oliver Laurel was killed on a lonely road two miles from Trenton at 3:30 A.M., February 14. By Washington's Birthday, the police had rounded up five suspects in Philadelphia: in alphabetical order, Hank, Joey, Red, Shorty and Tony. Under questioning, each of the men made four statements, three of which were true and one of which was false. It is known that one of these men killed Oliver Laurel (we have that assurance from Charlie the Canary), and the task is to identify the murderer. From the police records, we get the following transcript of statements:

Hank: (a) I did not kill Laurel.
      (b) I never owned a revolver in my life.
      (c) Red knows me.
      (d) I was in Philadelphia the night of February 14.

Joey: (a) I did not kill Laurel.
        (b) Red has never been in Trenton.
        (c) I never saw Shorty before.
        (d) Hank was in Philadelphia with me the night of February 14.

Red: (a) I did not kill Laurel.
        (b) I have never been in Trenton.
        (c) I never saw Hank before now.
        (d) Shorty lied when he said I am guilty.

Shorty: (a) I was in Mexico City when Laurel was murdered.
        (b) I never killed anyone.
        (c) Red is the guilty man.
        (d) Joey and I are friends.

Tony: (a) Hank lied when he said he never owned a revolver.
        (b) The murder was committed on St. Valentine's Day.
        (c) Shorty was in Mexico City at that time.
        (d) One of us is guilty.

To make up for our lack of acumen, which Sherlock Holmes apparently possessed, we amateurs shall rig up a kind of lie detector to help us put together all the pieces of the puzzle. Ours is simply a "sentence sorter" whose blueprint is based on Tables 1a and 1b.

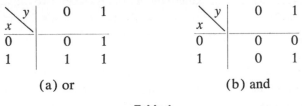

x \ y	0	1
0	0	1
1	1	1

(a) or

x \ y	0	1
0	0	0
1	0	1

(b) and

Table 1

Let's look into how sentences may be sorted. Take the simple sentence, "It is raining." If it is a true statement, we say that its *truth value* is "true"; if it is a false statement, we say that its truth value is "false." (It should not be both true and false simultaneously.) Now the statement, "The ground is wet" also has the two possible truth values: true and false; not both true and false simultaneously. These are simple sentences and may be combined to form compound sentences with the linking words *and* and *or*, such as:

> It is raining or the ground is wet.
> It is raining and the ground is wet.

The "or" sentence is called a disjunctive sentence, and the "and" sentence is called a conjunctive sentence. The natural question is now whether we can determine the truth value of these compound sentences if we know the truth value of each simple sentence, and the answer is Yes. For the statement, "It is raining or the ground is wet," if it is not raining and the ground is not wet, the statement is certainly false. If it is raining but the ground is not wet (the ground may be under a big canopy), is the compound sentence still true? It is. Also, if it is not raining, but the ground is wet (someone tossed a bucket of water), the compound statement remains true. Finally, if it is both raining and the ground is wet, the truth value is obviously true. Using 1 to stand for a true statement and 0 for a false statement, the result of the compound sentence involving the linking word "or" is shown in Table 1a, where the left-hand column shows the truth value of the first phrase x, the top row the truth value of the second phrase y, and the entry shows the result of the compound statement. To simplify matters we will use a plus sign ($+$) to denote the "or" disjunction. Since Table 1b reminds us of the usual multiplication table, we shall further borrow the mathematical notation of multiplication for the "and" conjunction.

We plan to grind the statements of the suspects, in the form of compound sentences, into such a table and learn the truth. Before doing this, it will be useful to notice a few things in this "arithmetic." Aside from the usual but important facts of algebra such as $x + y = y + x$, $xy = yx$, et cetera, a few unexpected results should be called to mind. One of these is the statement that if $xy = 1$, then both $x = 1$ and $y = 1$. This we have observed before, but since we shall make use of the fact extensively, we want to emphasize it again. Other unexpected results are that whatever x

and y may be, if the value of x agrees with the value of y, then $x + y$ has this same value; so does xy. This can be stated:

$$a + a = a \quad \text{and} \quad aa = a,$$

and can be verified by inspecting the tables or by trying out all possible values for $a (= 0, 1)$. The first is the property of simplicity for addition and the second the property of simplicity for multiplication. Another useful property to be noticed is that no matter what a is, $a + 1 = 1$. This is a special case of the property called *absorption*.

The system such as we have in Table 1 is an example of Boolean algebra named after the English mathematician George Boole (1815–1864), who developed the logic in 1854 in his book *An Investigation of the Laws of Thought*. Intensive investigation is going on nowadays in Boolean algebra, and new and important theorems are being evolved. For our purpose, we shall try to get by with a few basic theorems.

One key to our mystery is that of the four statements, three must be true and one must be false. As a consequence, for each individual, all four of his statements cannot be true at the same time. In the language that our machine can understand,

$$abcd = 0.$$

Also, out of the various combinations of three statements at a time, *abc, bcd, cda, dab,* one of these conjunctions will be true, while the other three will be false. (A false statement in any of these will make that combination false, but we are guaranteed that there are three which are true simultaneously.) So, another key to our sentence sorting is the relation $abc + bcd + cda + dab = 1$, for each suspect.

A third key is that for any individual, of any two of his statements, at least one is true; so, by the property of absorption, the disjunction of any two statements is true. For example, $a + b = 1$. We are now ready for business!

Starting with Tony's statements, we can immediately spot two true statements: b, *the murder was committed on Valentine's Day;* and d, *one of us is guilty.* This may be stated $b(T) = 1$ and $d(T)$

= 1. When it is clear whose statement is referred to, we may simply say, "Tony: $b = 1, d = 1$," or $Tb = 1, Td = 1$, for short. This leaves the job of screening out which of the two remaining statements is false.

Sometimes two statements may necessarily have the same truth value such as Ra and Rd, or Hd and Jd; these are *consistent* statements. Sometimes two statements may necessarily have different truth values such as Rd and Sc; these are *contradictory* statements. We shall make use of these as we proceed. It will be helpful in following the discussion to keep track of what we have uncovered as we go along. This we can do in the form of a box score, as in Table 2.

SUSPECT	a	b	c	d
HANK				
JOEY				
RED				
SHORTY				
TONY		t		t

Table 2

We enter t or f in the box for the statement of each of these gentlemen according to whether the statement has been found to be true or false. So far, the only certainty we have are t's in Tony's box for b and d.

Next, look at Red's statements, a and d. Either Red is guilty or not guilty. If Red is guilty, both of these are lies, and if Red is innocent, both of these are true. In other words, these are consistent statements. Since each individual is allowed only one falsehood, $ad = 1$; so $Ra = Rd = 1$, and we can enter t's in these boxes. To put it into our algebraic format:

a and d have the same truth value: $a = d$.
The disjunction of any two statements is true: $a + d = 1$.

Hence, $a + d = a + a = a = 1$;
Since $a = d, d = 1$.

Now Rd and Sc are contradictory statements, so $Rd = 1$ means $Sc = 0$. From the relation, $abc + bcd + cda + dab = 1$, since $Sc = 0$, the first three terms are zero and we are left with $dab = 1$; therefore, $a = b = d = 1$ for Shorty. Sa and Tc are consistent statements, so $Sa = 1$, means $Tc = 1$.

From the chart, we can see that by now Tony has his quota of three true statements; therefore the fourth, namely statement a, must be false. How does the machine analyze this? Remember the key: $abcd = 0$; since $b = c = d = 1$, we have $abcd = a1 = a = 0$. The remaining arithmetic may be indicated as follows (check the box scores):

Ta and Hb contradictory: $Ta = 0$ means $Hb = 1$.
Sd and Jc contradictory: $Sd = 1$ implies $Jc = 0$.
For Joe: $abc + bcd + cda + dab = 1$, $Jc = 0$ leaves $dab = 1$; so $a = b = d = 1$.
Jb and Rb consistent: $Jb = 1$ implies $Rb = 1$.
For Red: $abcd = 0$; $a = b = d = 1$ implies $abcd = c = 0$.
Rc and Hc contradictory: $Rc = 0$ implies $Hc = 1$.
Jd and Hd consistent: $Jd = 1$ implies $Hd = 1$.

Our chart now looks like this:

SUSPECT	a	b	c	d
HANK		t	t	t
JOEY	t	t	f	t
RED	t	t	f	t
SHORTY	t	t	f	t
TONY	f	t	t	t

Table 3

Finally, for Hank: $abcd = 0$, $b = c = d = 1$, so $abcd = a1 = a = 0$.

And with this, we have positively identified Lanky the Merciless, known in the professional circles as Lanky Hank. If we put to use more intricate theorems than those at our disposal, the results may be obtained in fewer steps without the need of picking the statements in the right sequence as we have done. Consistent statements can be coded with the same identifying letter or symbol, and contradictory statements can be coded in what are known as complementary symbols. Indeed, it is possible to incorporate all the information in a single equation and, with skill, read off partial answers in sophisticated use of Boolean algebra.

Now that we are heady with the success of catching one crook, we may want to guarantee success every time. To be this successful, we must insist that all our villains accept the crooks' honor system: that out of four statements to be made under the third-degree grilling, three must be true and one must be false. This is an unlikely condition, to be sure. But these methods do illustrate the power of systematic development of the algebra of logic.

7

NUMBER

BY POPULAR DEFINITION a mathematician is a fellow who is good at numbers. Most mathematicians demur. They point out that they have as much difficulty as anybody else in reconciling their bank statements, and they like to refer to supporting anecdotes, such as that Isaac Newton, who was Master of the Mint, employed a bookkeeper to do his sums. They observe further that slide rules and electronic computers were developed as crutches to help mathematicians.

All of this is obviously irrelevant. Who, if not the mathematician, is the custodian of the odd numbers and the even numbers, the square numbers and the round numbers? To what other authority shall we look for information and help on Fibonacci numbers, Liouville numbers, hypercomplex numbers and transfinite numbers? Let us make no mistake about it: mathematics is and always has been the numbers game par excellence. The great American mathematician G. D. Birkhoff once remarked that simple conundrums raised about the integers have been a source of revitalization for mathematics over the centuries.

Numbers are an indispensable tool of civilization, serving to whip its activities into some sort of order. In their most primitive application they serve as identification tags: telephone numbers, car licenses, ZIP-Code numbers. At this level we merely compare one number with another; the numbers are not subjected to arithmetical operations. (We would not expect to arrive at anything significant by adding the number of Leonard Bernstein's telephone to Elizabeth

Taylor's.) At a somewhat higher level we make use of the natural order of the positive integers: in taking a number for our turn at the meat counter or in listing the order of finish in a race. There is still no need to operate on the numbers; all we are interested in is whether one number is greater or less than another. Arithmetic in its full sense does not become relevant until the stage at which we ask the question: How many? It is then that we must face up to the complexities of addition, subtraction, multiplication, division, square roots and the more elaborate dealings with numbers.

The complexity of a civilization is mirrored in the complexity of its numbers. Twenty-five hundred years ago the Babylonians used simple integers to deal with the ownership of a few sheep, and simple arithmetic to record the motions of the planets. Today mathematical economists use matrix algebra to describe the interconnections of hundreds of industries, and physicists use transformations in "Hilbert space"—a number concept seven levels of abstraction higher than the positive integers—to predict quantum phenomena.

The number systems employed in mathematics can be divided into five principal stages, going from the simplest to the most complicated. They are (1) the system consisting of the positive integers only; (2) the next higher stage, comprising the positive and negative integers and zero; (3) the *rational numbers,* which include fractions as well as the integers; (4) the *real numbers,* which include the irrational numbers, such as π; (5) the *complex numbers,* which introduce the "imaginary" number $\sqrt{-1}$.

The positive integers are the numbers a child learns in counting. They are usually written 1, 2, 3, 4, . . . but they can and have been written in many other ways. The Romans wrote them I, II, III, IV, . . . ; the Greeks wrote them α, β, γ, δ, . . . ; in the binary number system, containing only the digits 0 and 1, the corresponding numbers are written as 1, 10, 11, 100, . . . All these variations come to the same thing: they use different symbols for entities whose meaning and order are uniformly understood.

Early man needed only the first few integers, but with the coming of civilization he had to invent higher and higher numbers. This advance did not come readily. As Bernard Shaw remarked in

Man and Superman, "To the Bushman who cannot count further than his fingers, eleven is an incalculable myriad." As late as the third century B.C. there appears to have been no systematic way of expressing large numbers. Archimedes then suggested a cumbersome method of naming them in his work *The Sand Reckoner.*

Yet, while struggling with the names of large numbers, the Greek mathematicians took the jump from the finite to the infinite. The jump is signified by the three little dots after the 4 in the series above. They indicate that there is an integer after 4 and another after the successor to 4, and so on through an unlimited number of integers. For the ancients this concept was a supreme act of the imagination, because it ran counter to all physical experience and to a philosophical belief that the universe must be finite. The bold notion of infinity opened up vast possibilities for mathematics, and it also created paradoxes. Its meaning has not been fully plumbed to this day.

Oddly, the step from the positive to the negative integers proved to be a more difficult one to make. Negative numbers seem altogether commonplace in our day, when 10 degrees below zero is a universally understood quantity and the youngest child is familiar with the countdown: ". . . five, four, three, two, one, . . ." But the Greeks dealt with negative numbers only in terms of algebraic expressions of the areas of squares and rectangles; for example

$$(a - b)^2 = a^2 - 2ab + b^2$$

(*see illustration on page 49*). Negative numbers were not fully incorporated into mathematics until the publication of Girolamo Cardano's *Ars Magna* in 1545.

Number Concepts can be arrayed in such a way that each succeeding system embraces all its predecessors. The most primitive concept, consisting of the positive integers alone, is succeeded by one extended to include zero and the negative integers. The next two additions are the rational and the irrational numbers, the latter recognizable by their infinitely nonrepetitive sequence of integers after the decimal. This completes the system of real numbers. The final array represents complex numbers, which began as a Renaissance flight of fancy and have since proved vital to the mathematics of physics and engineering. The complex numbers consist of real numbers combined with the quantity $\sqrt{-1}$, or i.

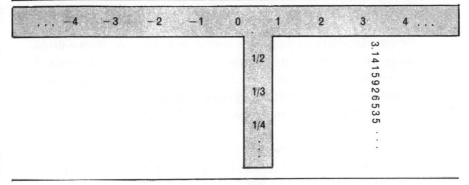

Fractions, or *rational numbers* (the name they go by in number theory), are more ancient than the negative numbers. They appear in the earliest mathematical writings and were discussed at some length as early as 1550 B.C. in the Rhind Papyrus of Egypt. The present way of writing fractions (for instance, $\frac{1}{4}$, $\frac{1}{5}$, $\frac{8}{13}$) and the present way of doing arithmetic with them date from the fifteenth and sixteenth centuries. Today most people probably could not be trusted to add $\frac{1}{4}$ and $\frac{1}{5}$ correctly. (Indeed, how often do they need to?) The handling of fractions, however, is by no means a dead issue. It recently became a matter of newspaper controversy as a result of the treatment of fractions in some of the new school mathematics courses, with the cancellation school pitted against the anticancellation school. The controversy stemmed from a divergence of opinion as to what the practical and aesthetic goals of school mathematics should be; the mystified layman, reading about it over his eggs and coffee, may have been left with the impression that everything he had once been taught about fractions was wrong or immoral.

The *irrational numbers* too have a long history. In the sixth century B.C. the mathematical school of Pythagoras encountered a number that could not be fitted into the category of either integers or fractions. This number, arrived at by the Pythagorean theorem, was $\sqrt{2}$: the length of the diagonal of a square (or the hypotenuse of a right triangle) whose sides are one unit long. The Greeks were greatly upset to find that $\sqrt{2}$ could not be expressed in terms of any number a/b in which a and b were integers, that is, any rational number. Since they originally thought the only numbers were rational numbers, this discovery was tantamount to finding that the diagonal of a square did not have a mathematical length! The Greeks resolved this paradox by thinking of numbers as lengths. This led to a program that inhibited the proper development of arithmetic and algebra, and Greek mathematics ran itself into a stone wall.

It took centuries of development and sophistication in mathematics to realize that the square root of two can be represented by putting three dots after the last calculated digit. Today we press the square-root button of a desk calculator and get the answer: $\sqrt{2} =$ 1.41421 .... Electronic computers have carried the specification of the digits out to thousands of decimal places. Any number that can be written in this form—with one or more integers to the left of a decimal point and an infinite sequence of integers to the right of the point—is a "real" number. We can express in this way the positive integers (for example, $17 = 17.0000 \ldots$), the negative integers ($-3 = -3.0000\ldots$) or the rational numbers ($17\frac{1}{5} = 17.20000\ldots$). Some rational numbers do not resolve themselves into a string of zeros at the right; for instance, the decimal expression of one-seventh is $1/7 = 0.142857\ 142857\ 142857\ldots$. What makes these numbers "rational" is the fact that they contain a pattern of digits to the right of the decimal point that repeats itself over and over. The numbers called "irrational" are those that, like the square root of 2, have an infinitely nonrepeating sequence of decimal digits. The best-known examples of irrationals are: $\sqrt{2} =$ 1.4142135623 ... and $\pi - 3.1415926535$ .... The irrational numbers are, of course, included among the real numbers.

It is in the domain of the *complex numbers* that we come to the numbers called "imaginary"—a term that today is a quaint relic of a more naïve, swashbuckling era in arithmetic. Complex numbers feature the "quantity" $\sqrt{-1}$, which, when multiplied by itself, produces -1. Since this defies the basic rule that the multiplication of two positive or negative numbers is positive, $\sqrt{-1}$ (or i, as it is usually written) is indeed an oddity: a number that cannot be called either positive or negative. "The imaginary numbers," wrote Gottfried Wilhelm von Leibniz in 1702, "are a wonderful flight of God's Spirit; they are almost an amphibian between being and not being."

From Renaissance times on, although mathematicians could not say what these fascinating imaginaries were, they used complex numbers (which have the general form $a + b\sqrt{-1}$) to solve equations and uncovered many beautiful identities. Abraham de Moivre discovered the formula

$$(\cos\theta + \sqrt{-1}\ \sin\theta)^n = \cos n\theta + \sqrt{-1}\ \sin n\theta.$$

	0	1	2	3	4	5	6	7	8	9	10	11	12	13	14	15	16
ABACUS PRINCIPLE																	
EGYPTIAN		I	II	III	IIII	∩ II	∩ II	≡	≡	≡≡	∩	∩I	∩II	∩III	∩IIII	∩≡	∩≡
MAYAN	•	•	••	•••	••••	—	•̱	•̱•	•̱••	•̱•••	═	•̇	•̇•	•̇••	•̇•••	≡	•̇̇
GREEK	A	B	Γ	Δ	E	F	Z	H	Θ	I	IA	IB	IΓ	IΔ	IE	IF	
ROMAN		I	II	III	IV	V	VI	VII	VIII	IX	X	XI	XII	XIII	XIV	XV	XVI
ARABIC	0	1	2	3	4	5	6	7	8	9	10	11	12	13	14	15	16
BINARY	00000	00001	00010	00011	00100	00101	00110	00111	01000	01001	01010	01011	01100	01101	01110	01111	10000

Leonhard Euler discovered the related formula

$$e^{\pi\sqrt{-1}} = -1$$

(*e* being the base of the "natural logarithms," 2.71828 . . .).

The complex numbers remained on the purely manipulative level until the nineteenth century, when mathematicians began to find concrete meanings for them. Caspar Wessel of Norway discovered a way to represent them geometrically (*see illustration on page 53*), and this became the basis of a structure of great beauty known as the theory of functions of a complex variable. Later the Irish mathematician William Rowan Hamilton developed an algebraic interpretation of complex numbers that represented each complex number by a pair of ordinary numbers. This idea helped to provide a foundation for the development of an axiomatic approach to algebra.

Meanwhile physicists found complex numbers useful in describing various physical phenomena. Such numbers began to enter into equations of electrostatics, hydrodynamics, aerodynamics, alternating-current electricity, diverse other forms of vibrating systems and eventually quantum mechanics. Today many of the productions of theoretical physics and engineering are written in the language of the complex-number system.

In the nineteenth century mathematicians invented several new number systems. Of these modern systems three are particularly noteworthy: quaternions, matrices and transfinite numbers.

Quaternions were Hamilton's great creation. For many years he brooded over the fact that the multiplication of complex numbers has a simple interpretation as the rotation of a plane. Could

Ancient and Modern Notations for the numerals from 1 to 16 are arrayed beneath the equivalent values set up on a two-rod abacus. Of the six examples all but two have a base of 10; these are repetitive above that number regardless of whether the symbol is tallylike or unique for each value. The Mayan notation has a base of 20 and is repetitive after the numeral 5. The binary system has a base of 2 and all its numbers are written with only a pair of symbols, 0 and 1. Thus two, or 2^1, is written 10; four, or 2^2, is written 100, and eight, or 2^3, is written 1000. Each additional power of 2 thereafter adds one more digit to the binary notation.

this idea be generalized? Would it be possible to invent a new kind of number and to define a new kind of multiplication such that a rotation of three-dimensional space would have a simple interpretation in terms of the multiplication? Hamilton called such a number a triplet; just as Wessel represented complex numbers by a point on a two-dimensional plane, the triplets were to be represented by a point in three-dimensional space.

The problem was a hard nut to crack. It was continually on Hamilton's mind, and his family worried over it with him. As he himself related, when he came down to breakfast one of his sons would ask: "Well, Papa, can you multiply triplets?" And Papa would answer dejectedly: "No, I can only add and subtract them."

One day in 1843, while he was walking with his wife along a canal in Dublin, Hamilton suddenly conceived a way to multiply triplets. He was so elated that he took out a penknife then and there and carved on Brougham Bridge the key to the problem, which certainly must have mystified passersby who read it:

$$i^2 = j^2 = k^2 = ijk = -1$$

The letters i, j and k represent hypercomplex numbers Hamilton called quaternions (the general form of a quaternion being $a + bi + cj + dk$, with a, b, c and d denoting real numbers). Just as the square of $\sqrt{-1}$ is -1, so $i^2 = -1$, $j^2 = -1$ and $k^2 = -1$. The key to the multiplication of quaternions, however, is that the commutative law does not hold (*see table on page 57*). Whereas in the case of ordinary numbers $ab = ba$, when quaternions are reversed, the product may be changed: for example, $ij = k$ but $ji = -k$.

The second modern number concept mentioned above, that of the matrix, was developed more or less simultaneously by Hamilton and the British mathematicians J. J. Sylvester and Arthur Cayley. A matrix can be regarded as a rectangular array of numbers. For example,

$$\begin{pmatrix} 1 & 6 & 7 \\ 2 & 0 & 4 \end{pmatrix}$$

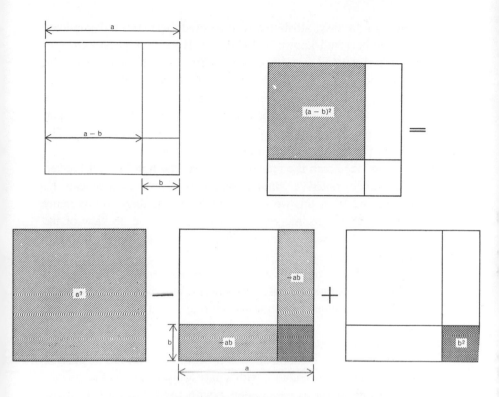

Negative Numbers were visualized by the Greeks in terms of lines and bounded areas. Thus they realized that the square erected on the line $a - b$ was equal in area, after a series of manipulations, to the square on the entire line a. The first manipulations require the subtraction of two rectangles of length a and width b from a^2. But these rectangles overlap, and one quantity has been subtracted twice. This is b^2, which is restored.

is a matrix. The entire array is thought of as an entity in its own right. Under the proper circumstances it is possible to define operations of addition, subtraction, multiplication and division for such entities. The result is a system of objects whose behavior is somewhat reminiscent of ordinary numbers and which is of great utility in many provinces of pure and applied mathematics.

The third modern concept that of transfinite numbers, represents a totally different order of idea. It is entertainingly illustrated by a fantasy, attributed to the noted German mathematician David Hilbert and known as "Hilbert's Hotel." A guest comes to Hilbert's Hotel and asks for a room. "Hm," says the manager. "We are all booked up, but that's not an unsolvable problem here; we can make space for you." He puts the new guest in room 1, moves the occupant of room 1 to room 2, the occupant of room 2 to room 3 and so on. The occupant of room N goes into room $N + 1$. The hotel simply has an infinite number of rooms.

How, then, can the manager say that the hotel is "all booked up"? Galileo noted a similar paradox. Every integer can be squared, and from this we might conclude that there are as many squares as there are integers. But how can this be, in view of the known fact that there are integers that are not squares, for instance, $2, 3, 5, 6 \ldots$?

One of the endlessly alluring aspects of mathematics is that its thorniest paradoxes have a way of blooming into beautiful theories. The nineteenth-century German mathematician Georg Cantor turned this paradox into a new number system and an arithmetic of infinite numbers.

He started by defining an infinite set as one that can be put into a one-to-one correspondence with a part of itself, just as the integers are in a one-to-one correspondence with their squares. He noted that every set that can be put into such correspondence with the set of all the integers must contain an infinite number of elements, and he designated this "number" as $\aleph$ (aleph, the first letter of the Hebrew alphabet). Cantor gave this "first transfinite cardinal" the subscript zero. He then went on to show that there is an infinity of other sets (for example the set of real numbers) that cannot be put into a one-to-one correspondence with the positive integers because they are larger than that set. Their sizes are represented by other transfinite cardinal numbers ($\aleph_1$, $\aleph_2$ and so on). From such raw materials Cantor developed an arithmetic covering both ordinary and transfinite numbers. In this arithmetic some of the ordinary rules are rejected, and we get strange equations such as $\aleph_0 + 1 = \aleph_0$. This expresses, in symbolic form, the hotel paradox.

The transfinite numbers have not yet found application outside mathematics itself. But within mathematics they have had considerable influence and have evoked much logical and philosophical speculation. Cantor's famous "continuum hypothesis" produced a legacy of unsolved problems that still occupy mathematicians. In recent years solutions to some of these problems have been achieved by Alfred Tarski of the University of California at Berkeley and Paul J. Cohen of Stanford University.

We have reviewed the subject matter (or dramatis personae) of the numbers game; it now behooves us to examine the rules of the game. To nonmathematicians this may seem to be an exercise in laboring the obvious. The geometry of Euclid is built on "self-evident" axioms, but rigorous examination of the axioms in the nineteenth century disclosed loopholes, inconsistencies and weaknesses that had to be repaired in order to place geometry on firmer foundations. But, one may ask, what is there about the simple rules of arithmetic and algebra that needs examination or proof? Shaken by the discoveries of the shortcomings of Euclid's axioms, and spurred by the surprising features of the new number concepts such as the quaternions, many mathematicians of the nineteenth century subjected the axioms of number theory to systematic study.

Are the laws of arithmetic independent, or can one be derived logically from another? Are they really fundamental, or could they be reduced to a more primitive, simpler and more elegant set of laws? Answers to questions such as these have been sought by the program of axiomatic inquiry, which is still going on. It has yielded rigorous and aesthetically appealing answers to some of them, and in the process it has brought forth new concepts such as *rings, fields, groups* and *lattices,* each with its own set of rules of operation and its own characteristic theory.

One of the major accomplishments, achieved in the 1870's, was the establishment of a set of axioms for the real numbers. It is summed up in the statement that the real-number system is a *complete ordered field.* Each of these words represents a group of rules that defines the behavior of the numbers.

First of all, the word "field" means a mathematical system in which addition and multiplication can be carried out in a way that

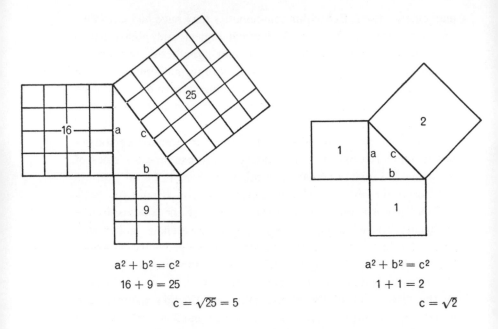

$$a^2 + b^2 = c^2 \qquad\qquad a^2 + b^2 = c^2$$
$$16 + 9 = 25 \qquad\qquad 1 + 1 = 2$$
$$c = \sqrt{25} = 5 \qquad\qquad c = \sqrt{2}$$

Irrational Numbers seemed paradoxical to the Greeks, who could not imagine numbers that were neither integers nor rational fractions but who could nonetheless express such numbers geometrically. In a right triangle with two sides of unit length 3 and 4 respectively, the hypotenuse is 5 units in length. But no rational fraction is equal to $\sqrt{2}$, the length of the hypotenuse of a right triangle that has sides of length 1. In effect, this says that an easily constructed line of quite tangible length is nonetheless "immeasurable."

satisfies the familiar rules, namely, (1) the commutative law of addition: $x + y = y + x;$ (2) the associative law of addition: $x + (y + z) = (x + y) + z;$ (3) the commutative law of multiplication: $xy = yx;$ (4) the associative law of multiplication $x(yz) = (xy)z;$ (5) the distributive law: $x(y + z) = xy + xz.$

Furthermore, a field must contain a zero element, 0, characterized by the property that $x + 0 = x$ for any element x. It contains a unit element, 1, that has the property that $1 \cdot x = x$. For any given element x of a field there is another element $-x$ such that

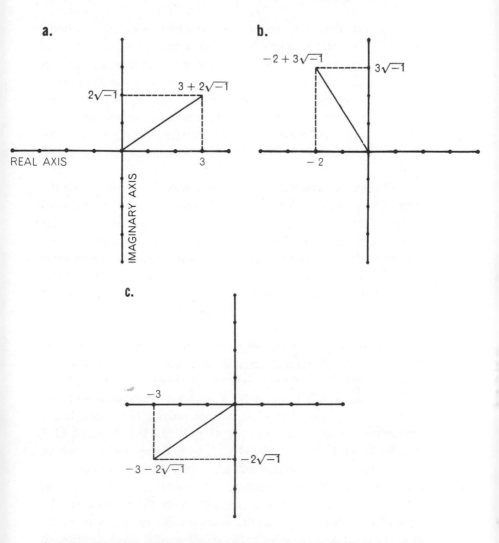

Complex Numbers can be represented and even manipulated in a geometric fashion. On the real, or *x*, axis each unit is 1 or −1. On the imaginary, or *y*, axis each unit is *i*, or $\sqrt{-1}$, or else −*i*. Thus all points on the
plane can be given complex numbers of the form *x* + *yi*. If a line through
both the origin and any point on the plane (*as shown in "a"*) is rotated
through 90 degrees (*as in "b"*), the result is the multiplication of the original complex number by $\sqrt{-1}$. A second rotation, and multiplication by *i*,
appears in "*c*."

$-x + x = 0$. This is the foundation on which subtraction is built. Another axiomatic property of a field is the cancellation rule of multiplication, that is, if $xy = xz$, then $y = z$ (provided that x is not equal to zero). Finally, for any element x (other than zero) a field must contain an element $1/x$ such that $x(1/x) = 1$. This is the basis for division. Briefly, then, a field is a system (exemplified by the rational numbers) whose elements can be added, subtracted, multiplied and divided under the familiar rules of arithmetic.

Considering now the second word, a field is "ordered" if the sizes of its elements can be compared. The shorthand symbol used to denote this property is the sign $>$, meaning "greater than." This symbol is required to obey its own set of rules, namely, (1) the trichotomy law: for any two elements x and y, exactly one of these three relations is true, $x > y$, $x = y$ or $y > x$; (2) the transitivity law: if $x > y$ and $y > z$, then $x > z$; (3) the law of addition: if $x > y$, then $x + z > y + z$; (4) the law of multiplication: if $x > y$ and $z > 0$, then $xz > yz$.

Finally, what do we mean by the word "complete" in describing the system of real numbers as a "complete ordered field"? This has to do with the problem raised by a number such as $\sqrt{2}$. Practically speaking, $\sqrt{2}$ is given by a sequence of rational numbers such as 1, 1.4, 1.41, 1.414 . . . that provide better and better approximations to it. That is to say, $1^2 = 1$, $(1.4)^2 = 1.96$, $(1.41)^2 = 1.9981$, $(1.414)^2 = 1.999396$ . . . Squaring these numbers yields a sequence of numbers that are getting closer and closer to 2. Notice, however, that the numbers in the original sequence (1, 1.4, 1.41 . . .) are also getting closer and closer to one another. We would like to think of $\sqrt{2}$ as the *limiting value* of such a sequence of approximations. In order to do so we need a precise notion of what is meant by the statement that the numbers of a sequence are getting closer and closer to one another, and we need a guarantee that our system of numbers is rich enough to provide us with a limiting number for such a sequence.

Following the path taken by Cantor, we consider a sequence of numbers in our ordered field. We shall say that the numbers of this sequence are getting closer and closer to one another if the difference of any two numbers sufficiently far out in the sequence is as small as we please. This means, for example, that all terms sufficiently far out differ from one another by at the most 1/10. If one wishes to go out still further, they can be made to differ by at most 1/100, and so forth. Such a sequence of numbers is called a "regular sequence." An ordered field is called a "complete" ordered field if, corresponding to any regular sequence of elements, there is an element of the field that the sequence approaches as a limiting value. This is the *law of completeness:* the "gaps" between the rational numbers have been completed, or filled up. It is the final axiomatic requirement for the real-number system.

All these rules may seem so elementary that they hardly need stating, let alone laborious analysis. The program of systematizing them, however, has been vastly rewarding. Years of polishing the axioms have reduced them to a form that is of high simplicity. The rules I have just enumerated have been found to be necessary, and sufficient, to do the job of describing and operating the real-number system; throw any one of them away and the system would not work. And, as I have said, the program of axiomatic inquiry has answered some fundamental questions about numbers and produced enormously fruitful new concepts.

The spirit of axiomatic inquiry pervades all modern mathematics; it has even percolated into the teaching of mathematics in high schools. A high-school teacher recently said to me: "In the old days the rules of procedure were buried in fine print and largely ignored in the classroom. Today the fine print has been parlayed into the main course. The student is in danger of knowing that $2 + 3 = 3 + 2$ by the commutative law but not knowing that the sum is 5." Of course anything can be overdone. Exclusive attention to axiomatics would be analogous to the preoccupation of a dance group that met every week and discussed choreography but never danced. What is wanted in mathematics, as in anything else, is a sound sense of proportion.

We have been considering how numbers operate; ultimately we must face the more elementary question: What *are* numbers, after all? Nowadays mathematicians are inclined to answer this question too in terms of axiomatics rather than in terms of epistemology or philosophy.

To explain, or better still to create, numbers it seems wise to try the method of synthesis instead of analysis. Suppose we start with primitive, meaningful elements and see if step by step we can build these elements up into something that corresponds to the system of real numbers.

As our primitive elements we can take the positive integers. They are a concrete aspect of the universe, in the form of the number of fingers on the human hand or whatever one chooses to count. As the nineteenth-century German mathematician Leopold Kronecker put it, the positive integers are the work of God and all the other types of numbers are the work of man. In the late nineteenth century, Giuseppe Peano of Italy provided a primitive description of the positive integers in terms of five axioms: (1) 1 is a positive integer; (2) every positive integer has a unique positive integer as its successor; (3) no positive integer has 1 as its successor; (4) distinct positive integers have distinct successors; (5) if a statement holds for the positive integer 1, and if, whenever it holds for a positive integer, it also holds for that integer's successor, then the statement holds for all positive integers. (This last axiom is the famous *principle of mathematical induction.*)

Now comes the *fiat lux* ("Let there be light") of the whole business. Axiom: There exists a Peano system. This stroke creates the positive integers, because the Peano system, or system of objects that fulfills the five requirements, is essentially equivalent to the set of positive integers. From Peano's five rules all the familiar features of the positive integers can be deduced.

Once we have the positive integers at our disposal to work with and to mold, we can go merrily on our way, as Kronecker suggested, and construct extensions of the number idea. By operations with the positive integers, for example, we can create the negative integers and zero. A convenient way to do this is by operating with *pairs* of positive integers. Think of a general pair denoted (a,b) from which we shall create an integer by the operation $a - b$. When a is greater than b, the subtraction $a - b$ produces a positive integer; when b is greater than a, the resulting $a - b$ integer is negative; when a is equal to b, then $a - b$ produces zero. Thus, pairs of positive integers can represent all the integers—positive, negative and zero. It is true that a certain ambiguity arises from the fact that a given integer can be represented by many different pairs; for instance, the pair (6,2) stands for 4, but so do (7,3), (8,4) and a host of other possible combinations. We can reduce the ambiguity to unimportance, however, simply by agreeing to consider all such pairs as being identical.

Multiplication Table for the quaternions, devised by William Rowan Hamilton, demonstrates the noncommutative nature of these imaginary quantities. For example, the row quantity *j*, multiplied by the column quantity *k*, produces *i*, but row *k* times column *j* produces −*i* instead. Each of the three quantities, multiplied by itself, is equal to −1.

	1	i	j	k
1	1	i	j	k
i	i	−1	k	−j
j	j	−k	−1	i
k	k	j	−i	−1

a.

$$\begin{pmatrix} a_1 & a_2 & a_3 \\ a_4 & a_5 & a_6 \\ a_7 & a_8 & a_9 \end{pmatrix} + \begin{pmatrix} b_1 & b_2 & b_3 \\ b_4 & b_5 & b_6 \\ b_7 & b_8 & b_9 \end{pmatrix} = \begin{pmatrix} a_1+b_1 & a_2+b_2 & a_3+b_3 \\ a_4+b_4 & a_5+b_5 & a_6+b_6 \\ a_7+b_7 & a_8+b_8 & a_9+b_9 \end{pmatrix}$$

$$\begin{pmatrix} 7 & 0 & 0 \\ -3 & 1 & -6 \\ 4 & 0 & 7 \end{pmatrix} + \begin{pmatrix} -8 & 0 & 1 \\ 4 & 5 & -1 \\ 0 & 3 & 0 \end{pmatrix} = \begin{pmatrix} 7-8 & 0+0 & 0+1 \\ -3+4 & 1+5 & -6-1 \\ 4+0 & 0+3 & 7+0 \end{pmatrix} = \begin{pmatrix} -1 & 0 & 1 \\ 1 & 6 & -7 \\ 4 & 3 & 7 \end{pmatrix}$$

b.

$$\begin{pmatrix} a_1 & a_2 & a_3 \\ a_4 & a_5 & a_6 \\ a_7 & a_8 & a_9 \end{pmatrix} \times \begin{pmatrix} b_1 & b_2 & b_3 \\ b_4 & b_5 & b_6 \\ b_7 & b_8 & b_9 \end{pmatrix} = \begin{pmatrix} a_1b_1+a_2b_2+a_3b_3 & a_1b_4+a_2b_5+a_3b_6 & a_1b_7+a_2b_8+a_3b_9 \\ a_4b_1+a_5b_2+a_6b_3 & a_4b_4+a_5b_5+a_6b_6 & a_4b_7+a_5b_8+a_6b_9 \\ a_7b_1+a_8b_2+a_9b_3 & a_7b_4+a_8b_5+a_9b_6 & a_7b_7+a_8b_8+a_9b_9 \end{pmatrix}$$

$$\begin{pmatrix} 6 & 0 & -1 \\ 1 & -3 & 2 \\ 8 & 5 & 6 \end{pmatrix} \times \begin{pmatrix} 4 & 2 & 3 \\ 0 & 1 & 6 \\ -5 & -1 & 7 \end{pmatrix} = \begin{pmatrix} 24+0+5 & 12+0+1 & 18+0-7 \\ 4+0-10 & 2-3-2 & 3-18+14 \\ 32+0-30 & 16+5-6 & 24+30+42 \end{pmatrix} = \begin{pmatrix} 29 & 13 & 11 \\ -6 & -3 & -1 \\ 2 & 15 & 96 \end{pmatrix}$$

Using only positive integers, we can write a rule that will determine when one pair is equal to another. The rule is that $(a,b) = (c,d)$ if, and only if, $a + d = b + c$. (Note that the later equation is a rephrasing of $a - b = c - d$, but it does not involve any negative integers, whereas the subtraction terms may.) It can easily be shown that this rule for deciding the equality of pairs of integers satisfies the three arithmetical laws governing equality, namely, (1) the reflexive law: $(a,b) = (a,b)$; (2) the symmetric law: if $(a,b) = (b,c)$, then $(b,c) = (a,b)$; (3) the transitive law: if $(a,b) = (c,d)$ and $(c,d) = (e,f)$, then $(a,b) = (e,f)$.

We can now proceed to introduce conventions defining the addition and the multiplication of pairs of positive integers, again using only positive terms. For addition we have $(a,b) + (c,d) = (a+c, b+d)$. Since (a,b) represents $a - b$ and (c,d) represents $c - d$, the addition here is $(a - b) + (c - d)$. Algebraically this is the same as $(a+c) - (b+d)$, and that is represented by the pair $(a+c, b+d)$ on the right side of the equation. Similarly, the multiplication of pairs of positive integers is defined by the formula $(a,b) \cdot (c,d) = (ac+bd, ad+bc)$. Here $(a,b)(c,d)$, or $(a - b)(c - d)$, can be expressed algebraically as $(ac+bd)$

Matrices are rectangular arrays of numbers, themselves without numerical value, that nonetheless can be treated as entities and thus can be added, subtracted, multiplied or divided in the proper circumstances. Such arrays offer a particularly convenient method for calculating simultaneous changes in a series of related variables. Addition is possible with any pair of matrices having the same number of columns and rows; row by row, each element in each column of the first matrix is added to the corresponding element in the corresponding column of the next, thus forming a new matrix. (The process is shown schematically at the top of the illustration and then repeated, with numerical values, directly below.) Multiplication is a more complex process, in which the two matrices need not be the same size, although they are in the illustration; a 3 × 2 matrix could multiply a 2 × 3 one. Each term in the upper row of the left matrix successively multiplies the corresponding term in the first column of the right matrix; the sum of these three multiplications is the number entered at column 1, row 1 of the product matrix. The upper row of the left matrix is now used in the same way with the second column of the right matrix to find a value for column 2, row 1 of the product matrix, and then multiplies the third column of the right matrix. The entire operation is repeated with each row of the left matrix.

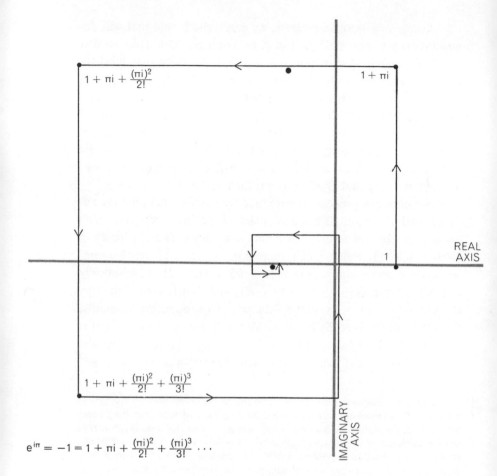

Virtuosity of complex numbers is demonstrated by conversion into geometrical form of the formula that relates e (the natural base of logarithms), π and √−1. The equation can be expressed as the sum of a series of vectors. When these are added and plotted on a complex plane, they form a spiral that strangles the point equal to −1.

− $(ad + bc)$, and this is represented on the right side of the equation by the pair $(ac + bd,\ ad + bc)$.

It can be shown in detail that all the familiar operations with integers (positive, negative and zero), when performed with such pairs of positive integers, will produce the same results.

Having constructed all the integers (as pairs of positive integers), we can go on to create all the other real numbers and even the complex numbers. The rational numbers, or fractions, which are pairs of integers in the ordinary system, can be represented as pairs of pairs of positive integers. For the real numbers, made up of infinite sequences of integers, we must set up infinite sequences of rationals rather than pairs. When we come to the complex numbers, we can again use pairs; indeed, it was for these numbers that the device of number pairs was first employed (by Hamilton). We can think of a complex number, $a + b\sqrt{-1}$, as essentially a pair of real numbers (a,b), with the first number of the pair representing the real element and the second representing the imaginary element of the complex number. Now, pairs will be considered equal only if they contain the same numbers in the same order; that is, $(a,b) = (c,d)$ only if $a = c$ and $b = d$. The rule for addition will be the same as in the case of the real numbers: $(a,b) + (c,d) = (a+c,\ b+d)$. This parallels the "ordinary" outcome of the addition of two complex numbers: $(a + b\sqrt{-1}) + (c + d\sqrt{-1}) = (a + c) + (b + d)\sqrt{-1}$. The multiplication formula for complex numbers, $(a,b) \cdot (c,d) = (ac - bd,\ ad + bc)$, also corresponds to the ordinary multiplication of such numbers: $(a + b\sqrt{-1})(c + d\sqrt{-1}) = (ac - bd) + (ad + bc)\sqrt{-1}$. Pairs of real numbers manipulated according to these rules reproduce all the familiar behavior of the complex numbers. And the mysterious $\sqrt{-1}$, that "amphibian between being and not being," emerges from the sea of axiomatics as the number pair $(0,1)$.

Thus, by four steps of construction and abstraction, we have advanced from the primitive positive integers to the complex numbers. Pairs of positive integers, combined in a certain way, lead to the set of all the integers. Pairs of integers (that is to say, now, pairs of pairs of positive integers), combined in a different way, lead to the rational numbers. Infinite sequences of rational numbers lead to real numbers. Finally, pairs of real numbers lead to the complex numbers.

Looking back over the 2,500 years that separate us from Pythagoras, we can make out two streams of thinking about numbers. There is the stream of synthesis, which began with tally marks and went on to build up number concepts of increasing complexity, in much the same way that a complex molecule is built up from atoms. On the other hand, there is a stream of analysis whereby mathematicians have sought to arrive at the essence of numbers by breaking down the complexities to their most primitive elements. Both streams are of enormous importance. Professional mathematicians today tend to play down number as such, favoring the qualitative aspects of their science and emphasizing the logical structure and symbolic potentialities of mathematics. Nevertheless, new ideas about number keep making their way into the mathematics journals, and the modern number theories are just now diffusing rapidly throughout our educational system, even down to the elementary schools. There are programs and committees for teaching advanced number concepts, from set theory to matrices, to students in high school. It seems safe to say that the coming generation will be imbued with an unprecedented interest in the fascinating uses and mysteries of numbers.

8

THE PHILADELPHIA STORY

A WHILE BACK, IN AN AFTER-DINNER SPEECH to a scientific society, I told about new applications of mathematics, *not* to the physical sciences, but to the social and economic sciences. Since wives were present at the talk and since I wanted them to be able to follow the thread of my argument for at least five minutes, I tried to find a very simple application that they could easily comprehend. Now what is simpler than 1, 2, 3, 4, 5, 6 . . . ? Where do they see it applied? In bake shops, in meat counters; wherever there is a pile up of people who must be served in turn.

Here is the excerpt from my address:

On a recent trip to New York, I got off at Pennsylvania Station. I walked to the taxi platform; but the train had been crowded, and soon dozens of people poured out of the station for cabs. Some of them waved and yelled; some stepped in front of the cabs rushing down the incline; some, who had porters in their employ, seemed to be getting preferential treatment. I waited on the curb, convinced that my gentle ways and the merits of my case would ultimately attract a driver. However, when they did, he was snatched out from under me. I gave up, took the subway, cursing the railway, the cabbies and people in general, and hoping they would all get stuck for hours in the crosstown traffic.

Several weeks later, on a trip to Philadelphia, I got off

at the 30th Street Station. I walked to the taxi platform. A sign advised me to take a number. A dispatcher loaded the cabs in numerical order, and I was soon on my way to the hotel. It was rapid; it was pleasant; it was civilized. And this is a fine, though exceedingly simple, way in which mathematics may affect social affairs. The criterion for loading is the order of arrival on the platform. The numbers have an order and can be used to simulate a queue without the inconvenience and indignity of actually forming a queue. The numbers are a catalyst that can help turn raving madmen into polite humans.

Today, people are very much concerned lest they allow themselves to be overregulated and overarithmetized. They are afraid of becoming a hole in a punched card and being processed like a number. Lest we worry about this too much, let me tell the sequel to the above true story.

I had not been in Philadelphia for many years. But one day last fall, I found myself at the 30th Street Station and needed a cab to get to the Philadelphia Airport. It was raining lightly, but steadily. I thought to myself, it's going to be a mess, getting a cab. Then I remembered the numbers on the platform. Well, maybe not.

When I got to the loading platform, the numbers were still there, all right, but their manner of operation was vastly different. If you wanted a Yellow Cab (or was it a Checker Cab?) you could take a number. Otherwise not. If you wanted an independent cab —and those seemed to be more plentiful than Yellows—you did not take a number; you pushed and shoved as in the old days. The traveler was now confronted with the delicious option of being civilized and taking a number, with the risk of arriving at his destination a half hour late (the rain was stronger and cabs were scarce), or ignoring the numbers and shoving one's way to punctuality. In fact, there was a third option, which combined the best features of both: take a number and then shove for an independent if it turned out advantageous.

Sweet chaos restored in Philadelphia!

9

POINSOT'S POINTS AND LINES

IN 1809 NAPOLEON invaded Austria and put Europe in a tur-
moil. That same year, a French mathematician by the name of
Louis Poinsot (1777–1859) spent much of his time quietly doodling
points and lines. While Europe was in flames, Poinsot calmly
traced stars and circles with compass and ruler. Poinsot could not
know that his idle speculations on the nature of polygons would
eventually help electronic computers to solve complex problems.

Let us look over Poinsot's shoulder as he draws his doodles.
He begins by drawing a number of points at evenly spaced intervals around the circumference of a circle. He then joins the points
with line segments. First he joins the points consecutively. Then he
joins every other point, then every third point, and so forth. Soon,
Poinsot's notebook is filled with stars and decorative designs of all
shapes.

We can easily retrace his doodles. Working with seven points,
and joining the points consecutively, we obtain Figure 1a, a regular heptagon. Figure 1b is obtained by skipping a point and joining
every second point. This leads to a seven-pointed star. Figure 1c
is obtained by joining every third point. This is also a seven-pointed
star, but of a different type from the previous one. Apparently,
there are two kinds of seven-pointed stars. Are there more? If we
join every fourth point, we are led back to Figure 1c. If we join
every fifth point, we obtain Figure 1b again. If we join every sixth

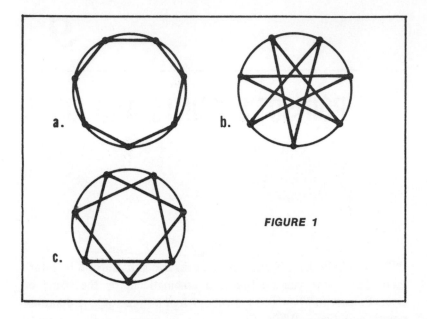

FIGURE 1

point, we come back to Figure 1a. The reason for this doubling up should be clear; the number 7 can be broken down into two-digit sums in only three ways: $4 + 3 = 7$, $5 + 2 = 7$, and $6 + 1 = 7$.

The stars in Figures 1b and 1c are regular polygons—not in the usual sense, but in a more general sense—for the sides have equal length and consecutive sides form equal angles.

Poinsot continued his doodles and found, to his surprise, that the situation with seven points was not typical. Following the plan of skipping a fixed number of points, one occasionally obtains a polygon with the same number of vertices as points started with. But it may also happen that one is led to a polygon with fewer vertices; some of the points cannot be reached. In this case we have a *degenerate* figure. This will happen, for example, if we start with eight points. Joining consecutive points, we obtain Figure 2a, a regular octagon. Joining every second point, the figure comes back on itself after a square has been traced out (*Figure 2b*). This square skips every other point. In order to accommodate the un-connected points into a starlike design, we must begin again, starting with an unconnected point. This leads to a second square, and the two overlapping squares form a design that is often seen on compass cards. By joining every third point, as in Figure 2c, we can connect every point and thus obtain a genuine polygon. When we join every fourth point, we no longer have a polygon, but only the degenerate Figure 2d.

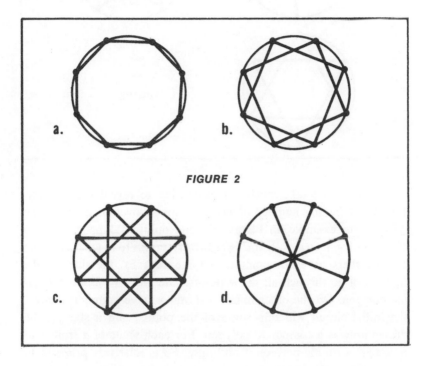

a.

b.

FIGURE 2

c.

d.

Some familiar designs can be generated in this fashion: the cross (*Figure 3a*), the five pointed star or pentagram (*Figure 3b*), and the star of David (*Figure 3c*). Figures 2b, 2d, 3a, and 3c, where degeneracy occurs, are known as *compound polygons*.

Poinsot was mystified by his results, and he set himself the problem of determining how many *n*-pointed regular polygons (noncompound) can be drawn. Call this number N. When $n = 7$, $N = 3$; when $n = 8$, $N = 2$. When $n = 9$, $N = 3$; when $n = 10$, $N = 2$. What is the general relationship between n and N?

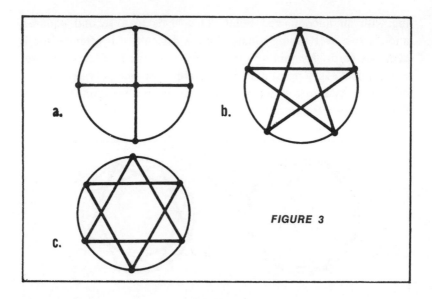

FIGURE 3

It does not take much experimenting to reveal the mystery. Let $d - 1$ be the number of points skipped over in drawing the design. For example, in Figure 3b, one point is skipped over, so that $d - 1 = 1$ or $d = 2$. If d is *relatively prime* to n (that is, if the two numbers have no common factor except 1), the sides of the figure will go through all the n points, and we will obtain a regular polygon. On the other hand, if d and n have common factors, the initial circuit will skip some of the points, and a star can be drawn only as a compound polygon. For each value of d from 1 to n, we get a simple polygon if and only if d is relatively prime to n. But since, as we observed, the figures are duplicated as we carry out all the possible skippings, N must be one half the number of integers less than n and relatively prime to n.

This is what Poinsot found, and he gave a formula for N. If the different prime factors of n are $a, b, c, \ldots e$, then

$$N = (n/2) (1 - a^{-1}) (1 - b^{-1}) (1 - c^{-1}) \ldots (1 - e^{-1}).$$

For example, if $n = 10$, its prime factors are 2 and 5, and

$$N = (^{10}\!/_2) (1 - 2^{-1}) (1 - 5^{-1}) = (^{10}\!/_2) (^{1}\!/_2) (^{4}\!/_5) = 2.$$

If n is itself prime, then $N = \frac{1}{2} (n - 1)$. Now that this much of the mystery of Poinsot's polygons has been revealed, the reader might like to amuse himself by discovering what happens when d and n have common factors.

I don't suppose that anyone would have predicted in 1809 that these figures, pretty as they are, would have amounted to much either in mathematics or in other sciences. Today, figures of this sort are of importance in a number of branches of applied mathematics. They have found use in electrical network theory, in statistical mechanics, and in numerical analysis. The writer has been concerned with the last-named application and would like to tell just a bit about it.

Numerical analysis is the branch of mathematics that studies the numerical solution of equations. The application derives from developing stars with an infinite number of vertices! Draw a circle with a compass. Readjust the opening of the compass, but make sure that it is not greater than the diameter of the circle. Place the compass point anywhere on the circumference, say at P_1, and allow the pencil to intersect the circumference at a second point P_2. Place the point on P_2 and allow the pencil to intersect the circumference at P_3. Proceed in this manner (*Figure 4*), always in one direction, either clockwise or counterclockwise. This yields a sequence of points P_1, P_2, P_3, . . . Draw the chords P_1P_2, P_2P_3, . . . The chords will be of equal length, and consecutive chords will make equal angles with one another. But will the later points of the sequence P_1, P_2, . . . ever come back and fall on top of an earlier point? If this happens, the process is cyclic, the polygon closes, and we obtain a Poinsot star. If it does not happen, we obtain a figure that is, so to speak, a regular polygon with an infinite number of sides.

Which of these types of figure is obtained will depend upon the relationship between the compass opening and the radius of the circle. If the angle subtended at the center of the circle by the chord P_1P_2 is a *rational* fraction of 360°, the figure will close. But if it is an irrational fraction of 360°, the figure will not close. In this case, the points go round and round the circle in what seems to be a helter-skelter fashion, spreading themselves over the whole circumference. If we take a look at an arc of the circumference,

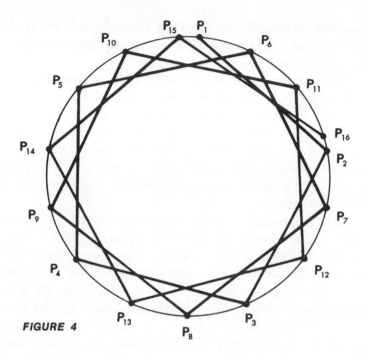

FIGURE 4

then eventually that arc—no matter how small it happens to be—will contain points of the sequence. In reality, the location of the points is not so random as it might appear. Every arc on the circumference receives its share of points in direct proportion to its length. This property is known as *equidistribution* and was demonstrated in 1916 by the famous mathematician Hermann Weyl (1885–1955).

The property of equidistribution possessed by infinite-sided Poinsot stars makes them particularly useful as an aid in carrying out certain very difficult computations on electronic machines.

10

CHAOS AND POLYGONS

FORM AND PATTERN have helped man to discover order in the chaos of the universe. The Pythagoreans saw order in the logic of numbers. These men were members of a secret society founded in the middle of the sixth century B.C. by the Greek philosopher Pythagoras. The society was perpetuated for more than one thousand years. In elevating number above all else, the Pythagoreans searched for patterns, such as those formed by pebbles in the sand. They formed triangles with pebbles for the "triangular" numbers (*see Figure 1*), rectangles for "rectangular" numbers, and squares for "square" numbers.

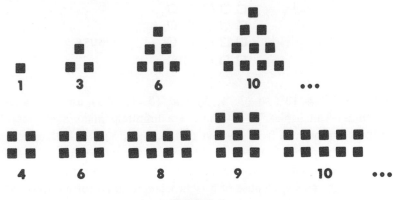

FIGURE 1

The triangular numbers are numbers of the form $\frac{1}{2}n(n+1)$; rectangular numbers are what we now call the composite numbers and are of the form mn; the (perfect) squares, of course, are the special cases of the rectangular numbers of the form $nn = n^2$ (m and n, positive integers).

Pebbles forming a particular triangle and pebbles forming the next larger triangle can be merged together to form a perfect square. This can be seen from the relation

$$\frac{1}{2}n(n+1) + \frac{1}{2}(n+1)(n+2) = \frac{1}{2}(n+1)(n+n+2)$$

$$= \frac{1}{2}(n+1)(2n+2) = (n+1)^2.$$

From this it is but a hop, skip and jump to the question of whether we can expect to merge the pebbles in two squares to re-group into a third square. That this can be accomplished can be seen in $3^2 + 4^2 = 5^2$ (*see Figure* 2),

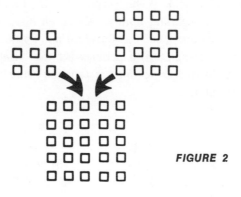

FIGURE 2

or $5^2 + 12^2 = 13^2$; so (3, 4, 5) and (5, 12, 13) are known as "Pythagorean triples." That there are infinitely many such possibilities can be seen from the fact that if (3, 4, 5) is a Pythagorean triple, then (6, 8, 10), (9, 12, 15), . . . are all Pythagorean triples. This is analogous to the fact that if (a, b, c) are the measures of the legs and hypotenuse of a right triangle, then for any positive number r, (ra, rb, rc) are also measures for the legs and hypotenuse of a right triangle. Of course, we restrict ourselves to positive integers for all the numbers involved. The proof of this theorem for both the Pythagorean triples and for the sides of the right triangle is easily seen; for, if (a, b, c) are numbers such that $a^2 + b^2 = c^2$, then

$$(ra)^2 + (rb)^2 = r^2a^2 + r^2b^2 = r^2 (a^2 + b^2) = r^2c^2 = (rc)^2.$$

As a result, each such triple and all the triples which can be obtained from it by multiplying by r are dumped into one equivalent class for study, and our attention is then turned to finding the possible representative sets of number triples.

That two squares cannot always be merged to form a third can be seen in $2^2 + 3^2 = 13$ or $2^2 + 2^2 = 8$. The simplest such example is $1^2 + 1^2 = 2$; in other words, it is impossible to build a square with two pebbles, having a whole number of pebbles on each side. In fact, it is not only impossible to find an integer x such that $x^2 = 2$; it is impossible to find even a number of the form $\frac{a}{b}$ (for whole numbers a and b) such that $\left(\frac{a}{b}\right)^2 = 2$. This fact was so disturbing to the Pythagoreans, who believed that any number must be obtainable from the whole numbers and the four basic arithmetic operations $+$, $-$, $\times$, $\div$, that they tried very hard to hide this discovery. It is thought that they either persuaded the discoverer to drown himself or threw him overboard in their attempt to keep this discovery from being known.

Eventually, geometry took a firm grip in Grecian culture. With it came questions of drawing figures meeting various requirements. The simplest is the line, for which we need only a straightedge. As soon as we have to produce a triangle with given line segments for the sides, we need a compass in addition to the straightedge. To construct a square, again the compass and straightedge will do; the compass being needed to erect a right angle.

Now, what can we do with just these two simple tools? There are some seemingly complicated constructions that can be drawn with them. For example, with precise enough instruments, we should be able to inscribe in a circle a regular polygon having 1024 sides if we so wish. Yet there are some terribly "simple" tasks that cannot be performed, such as trisecting an angle. Despite the tremendous efforts that have gone into the attempt, it has been proved (some hundred years ago) that in general an angle cannot be trisected with compass and straightedge alone. For certain special angles such as the right angle, this is easily done; in general, it is impossible—provided we keep to our agreement to limit our tools.

The problem of inscribing a regular polygon in a circle is the same as the problem of dividing a circle into n congruent parts or of cutting a pie into n pieces of equal size. If we find the point of division for one of the problems, we have the solution for the other. Cutting the pie in two is easy; we merely make a cut along one of the diameters. Since we can always bisect an angle (with compass and straightedge), we can halve the halves and repeat the process again and again. So a circle can always be divided into 2^n congruent parts. By the same token, if a circle can be split into r congruent parts by repeated subdividing, the circle can be divided into $2^n r$ congruent parts. Furthermore, if the length of a side of a regular polygon of n sides can be found entirely by the process of addition, subtraction, multiplication, division (by rational numbers), and the extraction of square roots, then the circle can be divided into n congruent parts—just by striking arcs with this length for radius.

Addition and subtraction of lengths can be done by laying off appropriate segments on a line; multiplication means just marking off segments end to end, and division can be done by marking off congruent segments on an auxiliary line (L' in Figure 3) and drawing parallels. The parallels intersect and divide the line into the proper number of segments. Extracting square roots is not as direct. For this, we need the Pythagorean theorem that the square

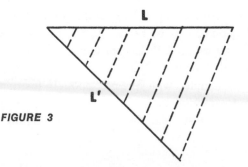

FIGURE 3

on the hypotenuse equals the sum of the squares on the sides. So, the square root of 2 is the hypotenuse of the triangle each of whose sides is 1 (*see Figure 4*). Having the $\sqrt{2}$, we can construct a right

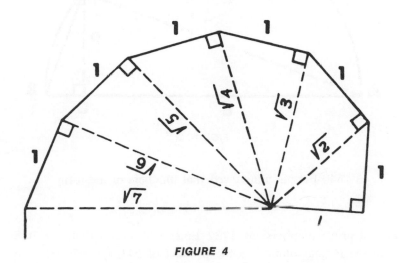

FIGURE 4

triangle with this segment as one side and for the other side, a segment of length 1. This process can be continued indefinitely in the form of the "spiral" started in Figure 4. However, this may not be satisfactory from a practical point of view because with each step we are likely to compound (physical) construction errors. To cut down the possible errors, we can do it most effectively by recalling a theorem in plane geometry about a perpendicular from the arc of a semicircle to its diameter. If arc ADB is a semicircle (*Figure 5*), and $\overline{DC}$ is perpendicular to $\overline{AB}$, the triangle ACD is similar to triangle DCB, and therefore the sides of the two triangles are in proportion. From this,

$$\frac{AC}{CD} = \frac{DC}{CB}, \text{ or } (AC)(CB) = (CD)^2. \text{ So } CD = \sqrt{(AC)(CB)}.$$

Thus, laying off $AC = 7$ and $CB = 1$, for example, we can get $CD = \sqrt{7}$.

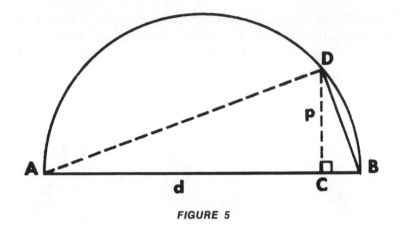

FIGURE 5

In 1640 Fermat announced that numbers of the form

$$2^{2^n} + 1$$

are all prime numbers; by 1732 Euler had shown that if n is 5 then the Fermat number is the product of 641 and 6700417, so the claim was disproved. Yet this does not destroy the fruitfulness of such numbers, for Karl Gauss (1777–1855) used these to prove that if p is a prime number of this form, then the circle can be divided into p congruent parts with compass and straightedge; furthermore, if a circle can be divided into a prime number of congruent parts, the prime number is necessarily of this form. The first five of these Fermat numbers (for $n = 0, 1, 2, 3, 4$) are 3, 5, 17, 257, 65537, and they are all primes; hence, from the assurance of Gauss, a circle can be divided into 3, 5, 17, . . . congruent parts.

Using the length of the radius to construct a regular hexagon is well known. By taking every other vertex, we have a trisection of the pie; and this quickly takes care of the case of three congruent parts. The case for 17 is not so easily discharged; Felix Klein expended eighteen pages of print in his *Famous Problems of Elementary Geometry* to this problem. In attending to the case for five congruent parts, it turns out that it is easier to shoot for ten first, and this is the approach that we shall take.

Since the central angle of each of the ten triangles in an in-

scribed regular decagon measures 36 degrees, each is an isosceles
72-36-72 triangle (*ABO* in Figure 6). Bisecting ∠ *ABO,* we have
triangle *ABC* also a 72-36-72 triangle.

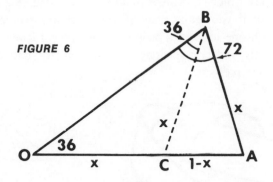

FIGURE 6

Therefore, the two triangles, *ABC* and *AOB* are similar. Also, since
both angles *CBO* and *COB* have the same measure of 36, triangle
CBO is isosceles, and $\overline{BC}$ is congruent to $\overline{OC}$.

If *AB* = *x,* then *CB* = *x* and *CO* = *x*.

If *AO* = 1, then *AC* = 1 − *x*.

From the similarity of triangles *ABC* and *AOB,* we have

$\frac{AB}{AO} = \frac{AC}{AB}$; that is, $\frac{x}{1} = \frac{1-x}{x}$, or $x^2 = 1 - x$.

Hence, the positive root is $\frac{1}{2}(\sqrt{5} - 1)$.

We can use the spiraling method described previously to construct $\sqrt{5}$, but since we already have a circle of radius 1, hence
of diameter 2, it will be more convenient to conceive of $\sqrt{5}$ as
$\sqrt{4+1} = \sqrt{2^2+1^2}$; that is, draw a tangent to the diameter at
A (*see Figure 7*) of length 1.

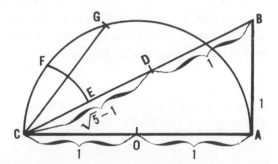

FIGURE 7

Then $BC = \sqrt{5}$. If $BD = 1$, then $CD = \sqrt{5} - 1$. Bisecting $\overline{CD}$ at E, we have the length of a side of the decagon, from which we can mark off arcs CF and FG to get $\overline{CG}$ as a side of the pentagon.

There are other equivalent methods of constructing a regular pentagon, but if you really want to construct one painlessly, you can do it with one loop of a strip of paper (*see Figure 8*). Draw the loop tightly and crease along the edges neatly, and there you have it, easy as pie!

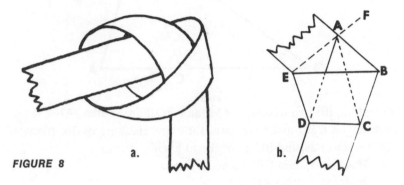

FIGURE 8

To see that you do have a regular pentagon, notice that by symmetry, $BC = ED = AB = DC = EA$. Therefore, angle EAD is congruent to angle EDA. Again by symmetry, angle BAC is congruent to angle EAD.

Since $\overline{DE}$ is parallel to $\overline{CA}$, angle CAD is congruent to angle EDA. We now have the three angles, EAD, CAD, and BAC congruent. If x is the measure of each of these angles, then $2x$ is the measure of angle FAB since line AD, unfolded, would lie along line AF. From this, we have $5x = 180$, and $x = 36$.

If x is 36, then the measure of angle BAE, or $3x$, must be 108 degrees, which is exactly what each of the inscribed angles of the regular pentagon should measure. Did you ever realize that each time you were tying a simple knot, you were constructing a regular pentagon?

Now that we have disposed of the construction of regular polygons of 3, 5 and 17 sides, the next move is yours—to construct polygons of 257 and 65,537 sides.

11

NUMBERS, POINT
AND COUNTERPOINT

IF YOU PLAY a musical instrument and have some knowledge of mathematics, you probably know that mathematics and music have much in common. The connection between these two seemingly unrelated subjects may appear in the study of harmony or counterpoint, or in the laws of vibrating strings and physical resonance. Although musicians use musical laws without analyzing them too deeply, a mathematical glance at musical structure may be quite illuminating.

We can start by examining the frequencies of vibrations for some of the tones. For a given frequency of vibrations giving off a particular tone, twice that frequency will emit the same tone one octave higher; thus 440 vibrations per second is the frequency for the A above middle C in what is called the "equally tempered" scale, and 880 vibrations per second is the frequency for the A that is one octave higher. In this scale, to which all keyed instruments have been tuned since the days of Bach (1685–1750), the intervals are equally tempered half tones. By this is meant that the ratio between frequencies for successive half tones is always the same (equal ratios).

Beginning with middle C, the half tones are:

$$C, C\sharp, D, D\sharp, E, F, F\sharp, G, G\sharp, A, A\sharp, B, \overline{C},$$

so that there are twelve intervals between octaves, an interval being determined by the ratio between the frequencies. If f is the frequency of C and r is the ratio between two adjacent half tones, then rf is the frequency of C$\sharp$, and r^2f is the frequency of D, etc.

$$
\begin{array}{cccccc}
C & C\sharp & D & D\sharp & \dots & \overline{C} \\
| & | & | & | & & | \\
f & rf & r^2f & r^3f & \dots & r^{12}f.
\end{array}
$$

Since the frequency of $\overline{C}$ is twice that of C,

$$r^{12}f = 2f, \text{ or } r = \sqrt[12]{2}\,.$$

This is why the ratio between successive half tones has been defined to be the 12th root of 2 in the equally tempered scale.

Normally, when we say that a sequence of numbers is equally spaced, we mean a sequence like: 1, 4, 7, 10, 13, 16, . . . , in which the *difference* between terms is constantly the same. Such a sequence is said to be in *arithmetic progression*. A sequence such as: 1, 4, 16, 64, 256, . . . , in which the *ratio* between terms remains constant is said to be in *geometric progression*. In view of this, we see that "equal intervals" in the equally tempered scale really means "equal" in the sense of a geometric progression.

The *diatonic,* or "scientific," scale is based on the same principle, with the exception that the number of vibrations per second for the C below the lowest A on the piano has been rounded off as 16 instead of 16.35, and similar rounding-off applies to frequencies of other tones in that octave. From this set of frequencies, the other octaves are obtained by doubling and repeated doublings. Eventually we come to 256 as the frequency of middle C. Frequencies for some of the other tones in the middle octave in this scale are:

C	D	E	F	G	A	B	$\overline{C}$
256	288	320	341⅓	384	426⅔	480	512.

Not all tones have been listed here, but even with a complete listing we do not expect the tones to be spaced equally in an arithmetic progression. Notice, however, the frequencies for C, E and G: 256, 320 and 384. These *are* in arithmetic progression, the difference between neighboring pairs being 64 vib./sec. These three tones make up what is called the major triad, or when played together (or in rapid succession) the major chord. A simple reduction will show that the ratio of the three frequencies is C:E:G = 4:5:6. Because of the simple relationship among these frequencies, some of the overtones merge and our ears pick out this set of tones, and we consider this combination to blend most harmoniously.

It has been discovered experimentally that on plucking strings that are of the same mass and stretched under equal tension, the shorter the string the higher is the tone emitted. This is why shortening the string of the violin, for example, by depressing it closer to the body of the instrument will result in a higher pitch. The formula showing this relationship is $nl = k$, where l represents the length, n the number of vibrations per second, and k is a constant (number) determined by the tension and the unit mass of the string. From this, we can see that, since by depressing the string in the middle we have effectively shortened the string in half, the number of vibrations per second would have doubled. So the string depressed in the middle would emit a tone one octave higher than the free string.

We can read into the equivalent formula, $l = k/n$, another interesting relationship. Noting the ratio of frequencies of C:E:G is 4:5:6, the ratio of the lengths of the string would be $\frac{1}{4}:\frac{1}{5}:\frac{1}{6}$ or, equivalently (multiplying by the least common denominator, 60), 15:12:10. What this may say is that if a 15-inch string has been tuned to C, then E would be sounded when the string is stopped at the 12-inch mark and G at the 10-inch mark. On the number line (*see Figure 1*), 0, 10, 12, and 15 is then considered a set of harmonic points; so is the set 0, 20, 24, 30, or 0, 5, 6, 7½, et cetera. The ratio of the lengths ¼, ⅕, ⅙ accounts for the name

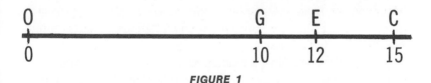

FIGURE 1

"harmonic sequence" that has been attached, logically enough, to: $\frac{1}{2}, \frac{1}{3}, \frac{1}{4}, \frac{1}{5}, \frac{1}{6}, \ldots \frac{1}{n}, \ldots$, although strictly speaking, the major triad is identified with just three of these terms.

The example of a harmonic set of points was drawn from the diatonic scale because of the simplicity in working with the approximate frequencies. In so doing, the ratio for the major triad came out nice and simple. From the standpoint of music appreciation, tones with these integral measures blend more harmoniously than measures according to the equal ratios of the 12th root of 2 in the equally tempered scale. The advantage in the equal scale is simply that it makes no difference where the starting point is; the sequence of half tones bears the same relation throughout from one half tone to the next. This makes for such an ideal arrangement in transposition into different keys that the small deviation from the pure blending of the diatonic tones may be disregarded. The small deviation may be exemplified by checking the ratios within both scales. The ratio of frequencies between E and C in the diatonic scale is 5:4 or 1.25:1. Referring to the table pairing the tone with $r^n f$ in the equally tempered scales, we see that

$$\frac{\text{frequency of E}}{\text{frequency of C}} = \frac{r^3 f}{f} = \frac{r^3}{1}.$$

Since r is the 12th root of 2,

$$r^3 = 2^{3/12} = 2^{1/4} \approx 1.26.$$

Therefore the ratio is approximately 1.26:1 as against 1.25:1. This represents less than 1 per cent deviation; hence for convenience, the use of the equal-tempered scale is accepted in instrumentation.

The integral measures in frequencies in the diatonic scale result in bringing together some of the overtones to give a harmonious blend. Aside from this, the set (0, 10, 12, 15) is considered a harmonic set from yet another point of view—that of projective geometry. In projective geometry where relations between configurations are studied under various projections, it is proved that for any four points on a line, a relation known as the *cross-ratio* of the points, remains unchanged under any projection. Attaching a posi-

FIGURE 2

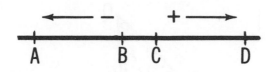

tive and negative sense to the line (*see Figure 2*) and using AB to mean the *directed distance* from A to B, we see that $AB = -BA$. In this notation, the cross-ratio of A, B, C, D, written $(ABCD)$, is by definition, $\dfrac{CA}{CB} \div \dfrac{DA}{DB}$.

We may illustrate what is meant by the remark that the cross-ratio remains unchanged under a projection by the diagram in Figure 3. If O is the center of projection, then any line l' will be cut by the lines of projection at corresponding points A', B', C', D', and the cross-ratio for these four points will be equal to the cross-ratio of the original four points; that is,

$$\frac{CA}{CB} \div \frac{DA}{DB} = \frac{C'A'}{C'B'} \div \frac{D'A'}{D'B'}$$

This can be proved by use of some trigonometry considering the areas of triangles OCA, OCB, ODA and ODB. When this cross-ratio is -1, the points are said to be a *harmonic set*. On this basis, $(OEGC)$ of Figure 1 is a harmonic set since $GO = -10$, $GE = 2$, $CO = -15$, $CE = -3$, and

$$\frac{GO}{GE} \div \frac{CO}{CE} = \frac{-10}{2} \div \frac{-15}{-3} = -1.$$

FIGURE 3

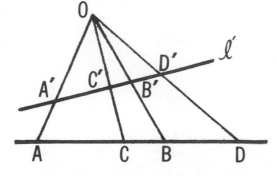

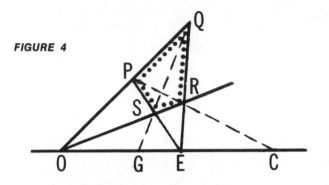

FIGURE 4

The set of points (O, E, G, C) is a special set which may be obtained from a quadrilateral $PQRS$ (*see Figure 4*) by extending the sides $\overline{PQ}, \overline{RS}, \overline{PS}, \overline{QR}$, and drawing the diagonals PR and QS. In projective geometry this figure is called a complete quadrangle. If

O is the intersection of PQ with RS,

E is the intersection of PS with QR,

and G is the intersection of QS with OE,

then C is the intersection of PR with OE.

This is proved by the tools of projective geometry, but can be verified in our instance thus: Draw any two lines through O; choose a point on one of the lines as Q (*see Figure 5*);

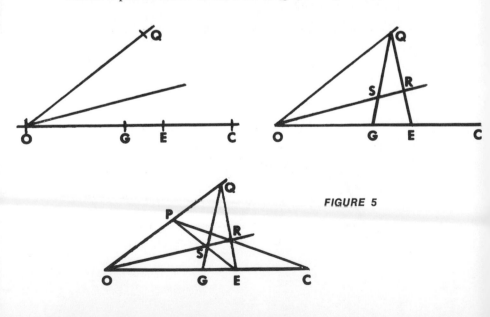

FIGURE 5

draw $\overline{QE}$ and $\overline{QG}$. Then the intersection of $\overline{QE}$ with the second line through O is R, the intersection of $\overline{QG}$ with $\overline{OR}$ is S, the intersection of $\overline{ES}$ with $\overline{OQ}$ is P, and it will be seen that $\overline{PR}$ does indeed intersect at C. In this way, (O, E, G, C) is a harmonic set in geometry as well as in music, and the major triad becomes a meeting ground for arithmetic, geometry, and music.

Since the cross-ratio depends on the order in which the points are named, for the same four points various cross-ratios may be formed. Apparently, the number of cross-ratios formed would be the same as the number of ways four objects may be permuted. For the first object, there are four choices; once a choice has been made for the first object, three choices remain for the second object; once these two choices have been made, two choices remain for the third object, et cetera. So, for all four objects there are $4 \times 3 \times 2 \times 1$ or 24 permutations. However, notice that

$$(ABCD) = \frac{CA}{CB} \div \frac{DA}{DB} = \frac{-AC}{-BC} \div \frac{-AD}{-BD}$$

$$= \frac{AC}{BC} \div \frac{AD}{BD} = \frac{AC}{AD} \div \frac{BC}{BD} = (CDAB)$$

In fact, $(ABCD) = (CDAB) = (BADC) = (DCBA)$, and in a similar manner all the 24 permutations may be separated into groups of four so that there are actually only six distinct groups of permutations, giving rise to six distinct cross-ratios.

By plowing through some arithmetic, it can be shown that if one of these cross-ratios is t, then the entire list of distinct cross-ratios for the four points is:

$$t, \frac{1}{t}, 1-t, \frac{1}{1-t}, \frac{t-1}{t}, \frac{t}{t-1}.$$

This is far from the end; the cross-ratio or *anharmonic* ratio, as it is sometimes called, plays a dominant role in projective geometry. Since straight lines remain straight lines under projection, and points of intersection remain points of intersection, it is reasonable from the tie-in of the cross-ratio with the parts of the quadrangle that the cross-ratio for a configuration will remain the same under whatever projection, and this is indeed the case.

Aside from the intuitively obvious places to look for the presence of the cross-ratios, we are pleasantly surprised now and then by their rather unexpected appearances elsewhere in mathematics. As a case in point, they turn up in the study of a basic, and at once important, concept in mathematics as elements of an abstract group.

The properties needed to define a *group* are few in number, but they form solid building blocks for the inner works of our machinery in arithmetic, algebra and other areas in mathematics. A set of elements, together with an operation, $\circ$, defined on them, constitutes a group if the following properties hold:

(i) if a and b are elements of the set S, then $a \circ b$ is an element of the set;

(ii) if a, b and c are elements of S, then $a \circ (b \circ c) = (a \circ b) \circ c$;

(iii) there is an element e in S such that $a \circ e = a$ for any element a of S;

(iv) for any element a of S, there is an element a' of S such that $a \circ a' = e$.

By trying a few samples, it can be seen that the set of integers would constitute a group under addition and the set of nonnegative rational numbers would constitute a group under multiplication.

Consider now each of the cross-ratios as an element of a set and define an operation, $\circ$, as follows:

By $a \circ b$ where a and b are cross-ratios, is meant that wherever t appears in a, t is to be replaced by the cross-ratio b.

For example,

$$(1 - t) \circ \frac{1}{t} \text{ means } \left(1 - \frac{1}{t}\right), \text{ which is equivalent to } \frac{t - 1}{t}.$$

Then it can be verified that the result of this operation will always be equivalent to one of the six given cross-ratios. Further checking will show that we have here indeed an example of a group.

The entire "times" table can be constructed without difficulty, although much demand would be made for close attention to details in the construction. Relabeling the cross-ratios according to the following scheme,

$$t = t, u = \frac{1}{t}, v = 1 - t, w = \frac{1}{1 - t}, x = \frac{t - 1}{t}, y = \frac{t}{t - 1},$$

we can arrive at Table I for the operation.

∘	t	u	v	w	x	y
t	t	u	v	w	x	y
u	u	t	w	v	y	x
v	v	x	t	y	u	w
w	w	y	u	x	t	v
x	x	v	y	t	w	u
y	y	w	x	u	v	t

Table I

As mentioned before, the requirements to be a group are simple, but the underlying concepts are far-reaching. If we understand these concepts we have gone a long way toward understanding what makes arithmetic tick. When we sit back and listen to the lilting tones of Tchaikovsky or the disciplined measures of Mozart, if our ears detect some elementary counterpoint we can be reassured that lurking in the shadows, subtly or manifestly, hovers the benign spirit of mathematics whose beauty may emerge in either arithmetic or geometry.

12

THE MATHEMATICAL BEAUTY CONTEST

WHAT IS YOUR candidate for the most beautiful curve? Is it the arch of a parabola, when seen against the sky as the cable of a great bridge? Is it perhaps the Spiral of Cornu, with its graceful reverse twist and the mystery of its infinite spiraling toward two points?

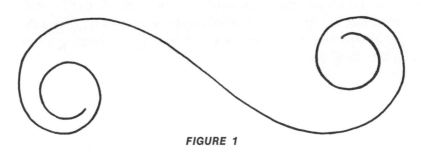

FIGURE 1

The Greeks, who held some rather impressive notions of beauty and perfection, came to the conclusion that the circle was the most beautiful curve. After all, the sun and moon were round, the horizon was round, and the planets went round in circles. The circle must be the perfect figure, for the architect of the universe would certainly not deal with imperfect creations. Some views about the circle were even more sweeping. The philosopher Empedocles held that the nature of God is a circle whose center is everywhere and whose circumference is nowhere.

But this theory received a blow even in ancient times. When astronomers were able to measure the paths of the planets accurately, they were found not to travel in true circles. In the second century, the astronomer Ptolemy, hard put for a simple description of what was happening, invented an ingenious system of cycles and epicycles—that is, of circles and circles upon circles—to generate the paths of the planets. In the seventeenth century, when Johannes Kepler found that the planets moved in ellipses around the sun, the circle was pretty much dethroned from its position as the most perfect of the curves.

I would like to raise the question once again. What is the most perfect curve? Notice that I've changed the question slightly, for I'm not going to talk about visual beauty, but conceptual beauty as it derives from the study of mathematics. Questions of beauty and perfection lie to some extent in the eye of the beholder. Therefore, I can only give you my answer, and you are free, after you have heard mine, to formulate your own. My answer is the Greeks' answer: the circle. I will defend my answer on the basis of the mathematical interest of the circle. I will do it by pointing out a number of remarkable things that are true of it.

Just as villains make the most exciting characters in novels and plays, a discussion of perfection tends to be dull unless it is relieved by constant contrast with what is imperfect. In considering the list of characteristics which follow, and which constitute my points of superiority for the circle, it is important to keep in mind the contrasting conditions of imperfection.

1. Every point on the circle is at the same distance from the center. This, of course, is the usual definition of a circle. It is the basis on which a circle is drawn with a compass or a circular chair leg turned on a lathe. This simple definition is why the circle is the easiest of all closed curves to draft accurately.

2. A closely related characteristic is that **if you spin a circle about its center, the rotated circle always occupies precisely the same space as did the original circle.** Each point on the circle bears the same relationship to its neighbors as does any other point to its neighbors. If this seems rather obvious, the strength of the statement can be felt by imagining what happens when a noncircular curve is rotated about a point. Think of elliptical wheels on a cart, or the plight of the billiard shark who was forced to play a match

> "On a cloth untrue,
> With a twisted cue
> And elliptical billiard balls."

Characteristics 1 and 2 lie at the heart of the operation of wheels, gears and rotating machinery of all sorts.

3. Every diameter of the circle is an axis of symmetry. This means that if you fold a circle over any diameter, the top half will fall exactly on the bottom half. Now the circle is certainly not the only figure that has an axis of symmetry. For instance, the square has four axes of symmetry: its two diagonals and the two lines through the center parallel to the edges. But for every line through a point to be an axis of symmetry is, I think, quite an achievement. Most figures have no axis of symmetry whatever.

4. A circle is a figure with constant width. This means that in all sets of two parallel lines tangent to the circle, the distance between the lines is constant.

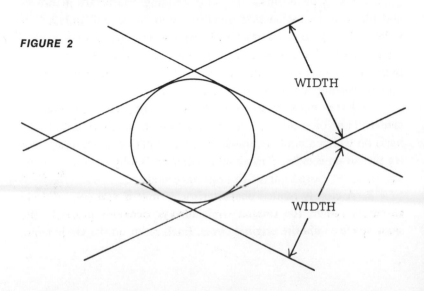

FIGURE 2

WIDTH

WIDTH

Can you imagine a figure other than a circle that has constant width no matter where it is measured? There are some, alas, for the circle must share this characteristic with certain other curves.

5. Every tangent to a circle is perpendicular to the radius drawn from the center to the point of tangency.

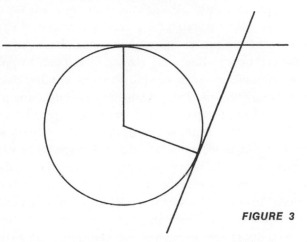

FIGURE 3

The converse is also true: If a curve is such that every line drawn from a fixed point to the curve is perpendicular to the tangent to the curve at that point, the curve must be a circle. This feature of the circle is important in *differential geometry,* which is a branch of geometry that studies the behavior of curves in the vicinity of individual points.

If you draw any old curve, then select a point not on it, and draw lines from the point to the curve, it will occasionally happen that one of the lines will be perpendicular to the tangent. For instance, if the curve is smooth, and if certain points on the curve are at a maximum or minimum distance to the given point, then it will occur there. Since all points on a circle are at a maximum (or, for that matter, minimum) distance from the center, all radii are perpendicular to the tangents.

6. Closely related is this property: **The curvature of a circle is constant at every point.** The converse is not completely true. If a curve has constant curvature, it must be either a circle or a straight line. (In mathematics, straight lines are also reckoned among the curves.)

The notion of the curvature of a curve is easily explained. Think of the curve as a road along which you are driving. Over the windshield of your car there is a compass which we will suppose acts instantaneously. At each point on the road, the compass tells you that you're traveling in a certain direction. (This is the direction of the tangent to the road at that point.) As you drive along, the compass reading may change from point to point. The curvature is defined as the amount the compass reading changes per unit of distance traveled. Thus, if you are traveling along a straight portion of road, the compass doesn't move, and hence the curvature is zero. On the other hand, if you travel along a hairpin turn, the compass will change violently, and the curvature will be very great.

I think it should not be difficult to show that a circle has constant curvature, and that the measure of its curvature is inversely proportional to the radius of the circle. The larger the radius, the smaller the curvature. From this point of view, a straight line, having (as it does) zero curvature, can be regarded as a circle with an infinite radius.

The statement that a circle has constant curvature is known in differential geometry as the *intrinsic equation of the circle*.

7. The last argument I will put forth in behalf of the circle is a most unusual one. It is a property of the circle that is not at all obvious by merely looking at it. **Of all the curves that enclose the same number of square inches of area, the circle is that curve that has the least perimeter.** This can be expressed in another way: **Of all the curves that have the same length of perimeter, the circle is the one that encloses the greatest area.** Think of it this way. Take a length of rope and tie it into a loop. Now put the loop on a table, and pull it around into different positions, trying to get the most area inside the loop. A circular loop is the one that does the trick.

This fact about circles can also be expressed with formulas. Draw any figure at all. Let its area be designated by A and its perimeter by L. Then

$$\frac{4\pi A}{L^2} \leq 1.$$

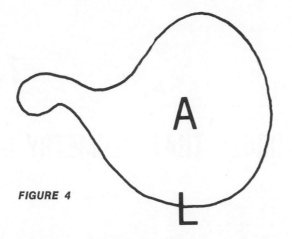

FIGURE 4

This theorem is the famous *isoperimetric inequality*. It is to be interpreted in the following way. The sign "$\leqq$" means "less than or equal to." The ratio

$$\frac{4\pi A}{L^2}$$

is always less than 1 if the figure is anything but a circle. If the figure is a circle, the ratio is precisely 1.

For a semicircle, this ratio is about .75, a value that is considerably less than 1. For a square, it is about .79, not too much better. For a regular hexagon, it is about .91. We can make the ratio closer and closer to 1 by making the figure resemble a circle more and more closely. But the top value of 1 is reserved for the circle alone.

My advocacy of the circle is now complete. I could have brought in many, many interesting and unique facts about the circle to strengthen my case, for hundreds are known, and (surprisingly enough) new ones both in elementary and in advanced mathematics are discovered yearly. But it would really be unnecessary. There is no figure which provides simplicity when simplicity is sought, and profundity when profundity is sought, after the manner of the circle. If we published a *Who's Who Among the Geometrical Figures,* and after the name of each figure listed its relations, achievements, distinctions, degrees, affiliations, and honors, the circle would have the longest list. There is no figure that can compare with it.

13

THE HOUSE THAT GEOMETRY BUILT

MOST OF US are aware that specialized language and vocabulary often go together with specialized branches of learning. Certain technical phrases—easily understood by the technically informed person—may entirely befuddle the uninitiated. As if to compound the confusion, we frequently employ abbreviations for pet phrases. For example, we use the innocent "etc." in everyday language; whether the reader reads it as "et cetera," "and so forth" or "and so on" really doesn't matter—all of these communicate the same message.

Geometry students are quite familiar with abbreviations such as "SAS," or "ASA," etc. "SAS" simply means that a certain pair of triangles may be proved congruent if "the sides and the included angle of one are correspondingly congruent to the sides and the included angle of the other." Another geometry favorite is "CPCFC," which seems like gibberish to those who have no idea that it means "corresponding parts of congruent figures are congruent," a familiar statement to the geometry student.

Thinking back to the activities in beginning geometry, you may remember that a good deal of effort went into proving the congruence of various figures. Why this emphasis on congruence?

The congruence of geometric figures allows us to study the properties of a particular figure with the understanding that any figure congruent to it will have the same properties. For example, if it can be proved in Figure 1, that the statement, "segment $\overline{AD}$ is congruent to $\overline{AE}$ and $\overline{BD}$ is congruent to $\overline{CE}$" implies that angle 1 is congruent to

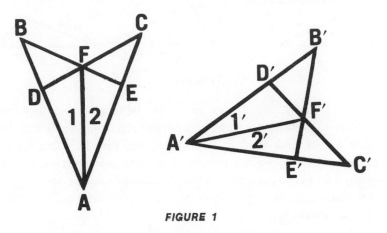

FIGURE 1

angle 2, then a corresponding statement may be made for the congruent figure A' B' F' C' A'. Without the understanding that corresponding properties apply to congruent figures, we may be faced with the task of repeating the study for each different position such a figure may occupy.

Since a figure is composed of points, and a point is an object that occupies a definite position in space, as soon as we move the model we are in reality studying a different set of points. That we can evolve the same discoveries is due to our tacit agreement that all properties of congruent figures are identical. Thus we have in essence sorted out all the possible congruences to a particular figure and dumped them all into one box. Congruent figures can then be considered equivalent just as $\frac{1}{2}$, $\frac{2}{4}$, $\frac{3}{6}$ are equivalent.

In practice, when we need to study the properties of any configuration, we simply make a copy of the figure and study it in the quiet of our "laboratory"—which may be three inches or three thousand miles from our original set of points. With this in mind, we must state explicitly what makes figures congruent.

At first thought, congruence can be defined by way of *super-*

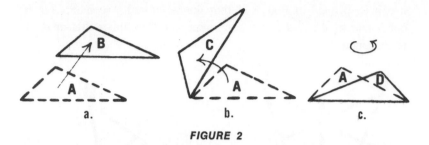

a.   b.   c.

FIGURE 2

position. Loosely, two figures are said to be congruent if they can be made to coincide by some appropriate rigid motion. But if the points are to have fixed positions, how can they be subjected to motion? Actually, we are merely visualizing moving a copy of a figure. The full statement of these conditions may be so burdensome that—once we understand the idea—they need not be stated over and over.

In Figure 2, we can see that the figures *B, C* and *D* are designed to be congruent to *A*. What is the "motion" that transforms *A* into each of the others? In Figure 2a it is just a shift, or *translation,* along the line indicated by the arrow; in Figure 2b it is a *rotation* in the plane about a certain point; and in Figure 2c it is a flipping over, or *reflection* of the plane, pivoted about a certain line. All the rigid motions possible are results of combinations of these three motions. For example, in Figure 3, from *A* to *B* is the result of a translation and a rotation; from *A* to *C* is the result of a translation and a reflection; from *A* to *D* is the result of a simple reflection or of a rotation and a reflection; and from *A* to *E* is the result of a translation, a rotation, and a reflection or of a simple reflection.

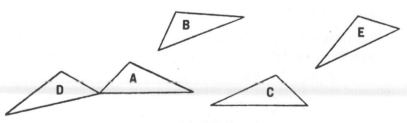

FIGURE 3

Allowing for the motion that leaves every point of the plane identically unchanged (the "identity" transformation), we can see by going through some of the motions, that the following set of four properties (principles) applies:

1. A translation followed by a translation is a translation (*see Figure 4a*).

2. In applying a chain of three translations it is immaterial whether the third translation is applied to the result of the first two translations or the result of the last two translations is applied to the first translation (*see Figure 4b*).

3. There is a translation that leaves every point unchanged.

4. For each translation there is a translation that transforms the configuration back to the original set of points (to each translation, there is an inverse—*see Figure 4c*).

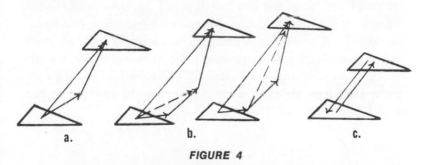

a.   b.   c.

FIGURE 4

The first of these statements is called the *closure* property; one translation followed by another does not result in "spilling over" outside the set of all translations. The second is the *associative* property; in a chain of translations the "grouping" of the translations is immaterial. Property 3 shows the existence of an *identity* translation and property 4 shows the existence of an *inverse* of each translation. These four properties appear so frequently in mathematics and are so important in the study of a mathematical system that a special name has been given to sets which may be so characterized. A set which possesses all four of these traits is said to form itself into a *group*—a very common word, used here in a technical sense.

You may be aware that the set of all integers forms a group under addition, and the set of all rational numbers ("fractions") forms a group under multiplication. In general, groups are quite prevalent in mathematical systems. The notion of a group can be further pursued for each of the other transformations included in our so-called rigid motions: It will be found that the set of all rotations in the plane constitutes a group; the set of all reflections about a line constitutes a group; the set of all rigid motions constitutes a group. Also, each of the sets, translation, rotation, and reflection, is a subgroup of the set of all rigid motions.

Are there groups of which the set of all these transformations is a subgroup? The answer is yes, and it pulls us into the orbits of other geometries. In *Euclidean* geometry where the transformations are the rigid motions, two figures are considered equivalent if they are congruent; here, the distance between any two points is unchanged by the transformation. In *projective* geometry, two figures are considered equivalent if one may be projected into the other. As the projections include *parallel* projections—in which the distances between points are unchanged—projective geometry includes Euclidean geometry as a special case. In addition to this, similar figures are equivalent (*see Figure 5a*).

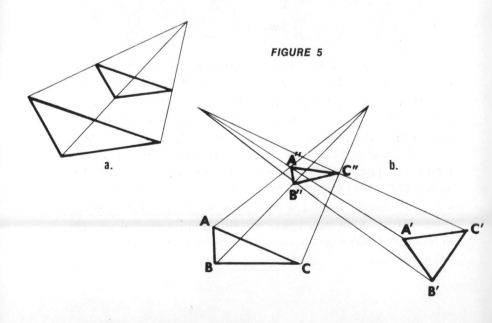

FIGURE 5

a.

b.

Furthermore, in projective geometry, *any* two triangles are equivalent. As an illustration, we can see this in Figure 5b:

Triangle ABC is equivalent to triangle $A''B''C''$ because one may be projected into the other;

triangle $A''B''C''$ is equivalent to triangle $A'B'C'$ for the same reason;

finally, triangle ABC is equivalent to triangle $A'B'C'$ through the equivalence of both to triangle $A''B''C''$.

So in projective geometry, we can show that all triangles may be placed in the same equivalence box (congruence box) and may be treated alike.

Because of rigidity, distance is invariant (unchanged) in Euclidean geometry. Distance no longer has to be invariant under projection, but if three points lie on a line in one configuration, the three points to which these have been transformed (their images) will likewise lie on a line; if three lines go through a single point, their images go through a single point, etc. In this way *collineation* (literally, "same line") is an invariant in projective geometry. Using coordinates, the transformations of Euclidean geometry may be expressed by polynomials of the first degree. In projective geometry, the transformations may be expressed as the ratio of polynomials of degree one. The class of all transformations which have an algebraic form gives rise to algebraic geometry, which includes projective geometry as a subset. All such transformations also constitute a group. Thus we have extension upon extension in geometry, each extension comprising all the others lower in the hierarchy. This systematized categorization of the geometries is due to the German mathematician Felix Klein and is known as his Erlanger Program.

Ultimately, we get to a branch of geometry in which extensive research is now being conducted. This is the branch known as *topology,* in which the intuitive concept may be presented thus: Imagine a configuration drawn on a rubber sheet. Any possible configuration which may be obtained after stretching, shrinking, bending, without ripping or tearing is equivalent to the original configuration. Thus the two figures in Figure 6a are in the same equiva-

FIGURE 6

lence box. This isn't all: under the terms set forth for "continuous deformations" in this study, the figures in 6b are also considered equivalent. A very general idea of sample equivalences in each of the geometries mentioned above may be indicated in a display as in Figure 7. This diagram is not meant to be all-inclusive, much is left for the imagination. But it does serve to indicate how one geometry encompasses others as subgroups, with topology as the great-granddaddy in this family.

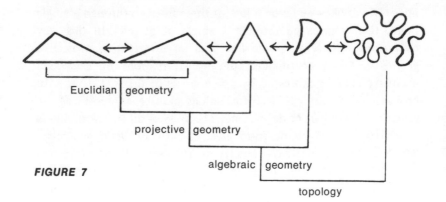

FIGURE 7

14

EXPLORERS OF THE Nth DIMENSION

HUGE ROCKETS soaring into space from launching pads in the United States and Russia have carried men a substantial number of miles in a substantially new direction—upwards.

In the first stage of flight, the rockets go almost straight up and travel many miles perpendicular to the surface of the earth. Thus, for the first time, three-dimensional motion on a large scale has been achieved.

Travel through outer space naturally turns our thoughts to the notion of dimension itself. What is dimension? How do we recognize that the world is three-dimensional? Are higher dimensions possible? How is dimension described mathematically? Can higher dimensions be visualized or experienced? Is it possible to travel through space in such a way that the dimensional character of space changes?

Much sense and nonsense have been spoken on these questions. Spiritualists talk of ghosts of the departed inhabiting a higher dimensional universe. Mystics claim that higher dimensions and different planes of reality can be reached through fasting, or through the use of exotic drugs. Physicists have suggested that time is a fourth dimension, and this is a notion that plays a central role in the space-time concept of the theory of relativity. But mathematicians, sitting at their desks, lifted off the launching pad and soared into space more than a century ago. They created multi-

101

dimensional universes on paper, and populated them with a variety of shapes and figures: hyperplanes, hyperspheres, polycylinders, tesseracts, and simplexes, to name a few.

The mathematical view of dimension is what concerns us here, as fascinating as other opinions might be. We shall explore it from two vantage points, the first geometric and the second algebraic. The mathematical world of higher dimensions is achieved through analogy and generalization. Length, breadth, and thickness are commonly called the three dimensions of space. On a plane surface (that is, a two-dimensional surface), there is no thickness, only length and breadth. Another way of putting it is that in a plane, we can find two lines that are mutually perpendicular, but we cannot find three lines that are mutually perpendicular. To find three such lines, we must step out into space. In space itself, we can find three mutually perpendicular lines, but we cannot, try as we may, find four such lines. To find four, we would have to go into the fourth dimension.

This state of affairs is illustrated by Figure 1. Figure 1a shows two axes at right angles. Figure 1b depicts three axes in space at right angles. What we have in Figure 1b is not the true space situation, but its projection onto a plane. By the same token, we can construct four axes in three-dimensional space and say that they are a projection of four mutually perpendicular lines in four-dimensional space. We can then show a projection of this three-dimensional figure on our two-dimensional paper. This is what we have done in Figure 1c.

Let's begin again. If a line is moved parallel to itself along a

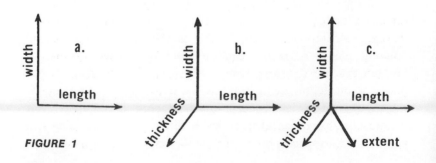

FIGURE 1

perpendicular axis, it traces out a square, as in Figure 2a. Similarly, if a square is moved parallel to itself along a perpendicular axis, it traces out a cube, as shown in Figure 2b. Or rather, what is shown is not the three-dimensional cube, but its projection. With a bit of imagination, we can go to the next higher stage. We take a three-dimensional cube and move it parallel to itself along a perpendicular axis in the fourth dimension (whatever this process means) and in this way generate a four-dimensional cube, or a *tesseract*. This is shown carried out in Figure 2c. The initial and final positions of the cube are drawn in heavy lines, while the other edges of the tesseract along which the cube has moved are dotted.

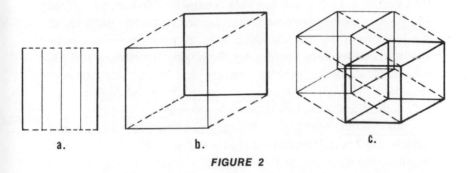

a.   b.
FIGURE 2
                                            c.

We can carry this process further. Move the tesseract along a new perpendicular axis and it will trace out a five-dimensional hypercube. We can draw the projection of such a figure on a plane, but it would be quite complicated.

Enough for the square. Let's take a look at the triangle. To draw a triangle, we take three points and connect the points two by two, as in Figure 3a. This procedure generalizes to the tetrahedron in three-dimensional space, obtained by taking four points not all in one plane and connecting them two by two, as in Figure 3b. What's to prevent us now from taking five points that don't lie in the same three-dimensional space and joining the points two by two? Such a figure is called a four-dimensional simplex, and its projection is shown in Figure 3c.

The algebraic approach to higher dimensions begins with an

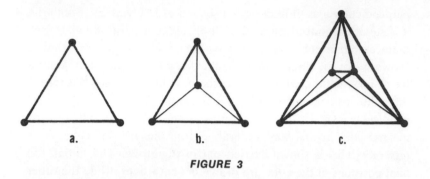

a.   b.   c.

FIGURE 3

observation of the mathematician René Descartes (1596–1650) that a point in a plane can be characterized by two numbers. These *coordinates,* as they are called, measure the respective distances of the point from two perpendicular axes. A point in space can be characterized by three coordinates. Following the obvious pattern, why shouldn't we say that four numbers characterize—or even constitute—a point in four-dimensional space? Why shouldn't n numbers constitute a point in n-dimensional space?

In two dimensions, a point is a pair of numbers (x,y) and a straight line is a relationship between x and y, such as $3x - 7y = 1$. In three dimensions, a point is a triple of numbers (x,y,z) and a plane is a relationship such as $3x + 6y + 4z = 5$. Again the pattern should be obvious; in n dimensions, a point is an n-tuple of numbers $(x, y, z, w, \ldots , t)$, and a relation between these numbers such as

$$x + 2y - z + 3w + \ldots + 7t = 1$$

must constitute a hyperplane. Two lines intersect in a point, and the algebraic counterpart of this is the fact that two first-degree equations in two variables generally have a single solution. In this way, algebraic interpretations of geometric facts can be built up in any number of dimensions. Algebra, in a brilliant stroke, creates n-dimensional geometry as a commonplace part of its day-to-day business, but reduces it to the mere study of equations and their solutions.

The algebraic method has a number of advantages over the geometric method. The ability of most people to generalize geometric figures is limited; a study of Figure 2c will attest to this fact. Algebra is able to reduce many spatial difficulties to near triviality. Here is an additional example: According to the French mathematician Henri Poincaré (1854–1912), the dimensional essence of space is that it can be separated into two parts by a figure of one less dimension. If a line is drawn in a plane, the plane is divided into two parts, and one cannot pass from one part to the other without crossing the line. The algebraic counterpart of this is the simple fact that the equation $ax + by + c = 0$ represents a line, and the two separated parts of the plane are represented by the inequalities.

$$ax + by + c < 0$$

and

$$ax + by + c > 0.$$

But even algebra has its limitations when it comes to the peculiarities of figures of higher dimension, and students of space are forced, from time to time, to augment the algebra with more pictorial representations. A four-dimensional hypersphere of radius 1 is to algebra the set of all points (x,y,z,w) that satisfy the equation

$$x^2 + y^2 + z^2 + w^2 = 1.$$

Its interior are those points that satisfy

$$x^2 + y^2 + z^2 + w^2 < 1,$$

while its exterior are those points that satisfy the reverse inequality.

This may satisfy some people, but not Professor Stefan Bergman of Stanford University, whose research on four dimensions and its relationship to complex analytic functions is world renowned. Bergman likes to picture a hypersphere as a moving picture of a three-dimensional sphere that grows and then collapses, as in Figure 4. Each frame is taken at a slightly later time, and in this way, time is used in this representation as a fourth dimension. By imagining two such figures, he allows them to intersect, and is able to study the various 3-, 2-, and 1-dimensional boundaries that have played an important role in his theory. I once asked Bergman why I had difficulty visualizing a four-dimensional situation that was obvious to him. He replied that he had been practising the art for years, and that practice makes perfect even in higher dimensions.

The impatient and perhaps outraged reader may now be saying: "Yes, yes, it is all very well to move a cube along a fourth

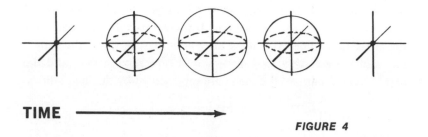

TIME ───────────►

FIGURE 4

dimension that doesn't exist and call the result a tesseract, or to throw a bunch of numbers together within two parentheses and call the result a point in higher space. But these things are not physical creations, and there is no reality to them. It is very much like a dictionary that gives a precise definition of a unicorn even though there are no unicorns."

This is admittedly the case. The mathematical speculations do not take us to a door and invite us to step beyond into a new physical universe of multidimensions. And yet, to deny reality to the concepts of the mind seems a bit severe. Consider the unicorn; he is found in many books, on tapestries, and on seals; a child could recognize a unicorn in the street and distinguish it from a lion. How much more could be accomplished if there "really" were such an animal?

If making multidimensional figures more "real" requires some kind of experience that relates to touching them and moving them around in space, then such experiences may not be far away. Such, at least, is the opinion of the distinguished Los Alamos mathematician S. M. Ulam, who has proposed the simulation of higher-space experiences by means of electronic computers. The mathematics of motion is not complicated. Having set up a four-dimensional object in a computing machine as a set of inequalities, we could twist some dials and move the object around, and at each stage of its motion have a projection of its three-dimensional sections displayed. It could be easily arranged that if two four-dimensional objects come into contact, a signal is given by the computing machine. In this way one could, for instance, learn to thread a four-dimensional object through a closed two-dimensional surface! This is an action for which there is certainly no visual or tactile experience. But the hand, together with the eye and the ear, coupled with the computer, would be provided with a set of inputs and outputs which the brain might fuse together as a genuine four-dimensional experience.

There are fat books that describe some of the things to be found in n-dimensional and even infinite-dimensional spaces. The dimensional explorers of the future may be found at the consoles of computing machines adding their bit to the catalog of the wonders of the universe.

15

THE BAND-AID PRINCIPLE

IT IS ALWAYS a delight when simple ideas lead to great consequences. Newton's apple led to the law of the mutual attraction of bodies, and the consequences of this can not be set forth adequately in fifty volumes. The story that I'm going to tell here is not so well known as Newton's apple. Its consequences are not so great. But the story is easier to understand: all that is required is a bit of geometrical and mechanical intuition. And it is a fine illustration of how mathematics tries constantly to understand and explain a complicated situation by an appeal to a related simple situation.

The shortest distance between two points, said the ancient geometers, is along a straight line. This is a simple enough statement to begin with, and one that is intuitively obvious. In Figure 1, the artist seems to be testing out this proposition to make sure that it is correct.

When we come to put this principle into practice, complications intervene. It may be applied when we take a short cut across a field, but in going, say, from New York to Tokyo, we are not likely to go in a straight line, unless some kind government provides us with a tunnel through the crust of the earth. A meaningful question that relates to travel on the surface of the earth is: given two points A and B on the surface of a sphere, what curve joining A to B, *and lying entirely on the surface of the sphere,* has the least length? The answer, as you may know, is an arc of the great circle that joins A to B. Great circles are circles that lie on the surface of the sphere and whose centers coincide with the center of the sphere.

But a more general question than this can be formulated, for while a sphere is undoubtedly of great importance to us Earth dwellers, it is certainly not the only curved surface that mathematicians have thought about. Given a surface: a cylinder, a cone, a sphere, an ellipsoid, a piece of tin that has been twisted into a

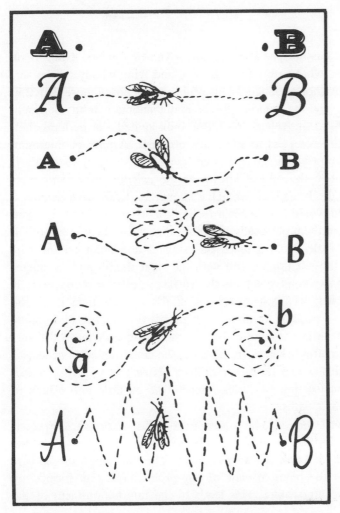

FIGURE 1

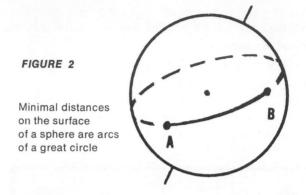

FIGURE 2

Minimal distances
on the surface
of a sphere are arcs
of a great circle

shape that has no special name, and given two points A and B on the surface, what path joining B to A and lying wholly on the surface has the shortest length? This problem was posed more than 250 years ago by the famous Swiss mathematician John Bernoulli (1667–1748). In August 1697, Bernoulli invited—or perhaps challenged—all geometers to solve this problem. A path of minimum length is known as a *geodesic,* and today, the study of geodesics is carried out in the subject known as *the calculus of variations* as well as in the branch of geometry known as *differential geometry.*

If you would like to get a feeling for what geodesics are, you can think of them physically. Suppose that the surface is a football or fancy bottle. Hold a piece of string at one point on the surface with the left thumb, and with the right thumb and forefinger located at a second point on the surface, pull the string taut. If the string lies on a convex portion of the surface, then it will be pulled into a position of minimal length between its two end points. If the surface is not wholly convex, the string may leave the surface and go off into space. In this case, the construction of geodesics with a string would be rather difficult. Now this is a mechanical way of constructing geodesics, and it tells us little about how to compute their paths.

Toward this end, let's acquire a bit of experience by considering some simple surfaces. The simplest surface of all is the plane, and we know the answer there. The next most simple surface (from the point of view of our problem) is that formed by two planes intersecting. This leads to the famous problem of the

FIGURE 3

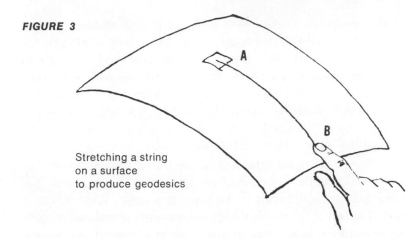

Stretching a string
on a surface
to produce geodesics

spider and the fly. A spider at point *A* on the west wall of a room spies a fly at point *B* on the north wall. The spider, who is mathematically inclined and does not like to build webs, would like to determine the shortest path along the walls to the fly. The fly, who is also mathematically inclined, is far more interested in seeing the problem solved than in preserving its identity. The problem can be solved by reducing it to that of two points in a single plane. Think

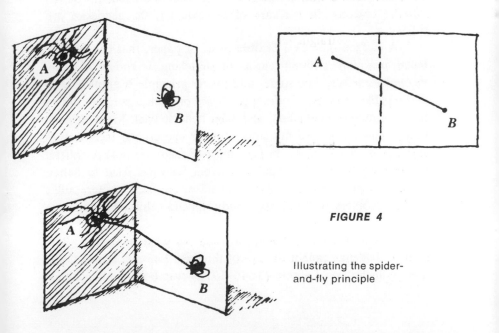

FIGURE 4

Illustrating the spider-
and-fly principle

of the two perpendicular walls as being hinged together. Flatten them out so that they form one plane. Then draw a straight line joining B to A. Fold the walls up again, and the geodesic is drawn in proper position. To prove this statement, suppose that some other path along the walls were shorter. When the walls are flattened to one plane, we would then have a path joining A and B which is not a straight line, but whose length is shorter than the straight line path. This would be impossible.

Now, it does not take a great stretch of the imagination to go beyond this; the principle of the spider and the fly can be applied more generally than to the walls of a room. What makes it tick? The fact that we can flatten out the surface without stretching or tearing it in any way so that it lies in a plane. The principle is therefore applicable to such surfaces. In addition to surfaces formed of sections of planes joined together, there are two common surfaces with this property: the cylinder and the cone. In order to find geodesics on a cylinder or a cone, we "unwrap" the surface, flattening it to a plane, draw the relevant straight line in the plane, and then wrap the plane up again. Using this idea, it is not too difficult to show that geodesics on a cylinder or on a cone are curves that make a constant angle with the elements of the cylinder or cone. In the case of the cylinder, the geodesics are helices.

A surface such as a flexible piece of paper, that can be flattened to a plane without tearing or stretching, is known as a *developable surface*. The spider and the fly principle is applicable to developable surfaces. To draw geodesics on such a surface, flatten it out, draw a straight line, and then bend it back to its original shape. This would end my story, but for one crucial fact: not all surfaces are developable; in fact, most of them are not. A sphere, for example, is not developable. Anyone who has tried to flatten out an orange peel knows this. And so we are back to Bernoulli's original problem without, apparently, having achieved much success.

Let's look back at the record and see what happened. Bernoulli's challenge caught the eye of the great Leibniz. Baron Gottfried Wilhelm von Leibniz (1646–1716) was both a philosopher

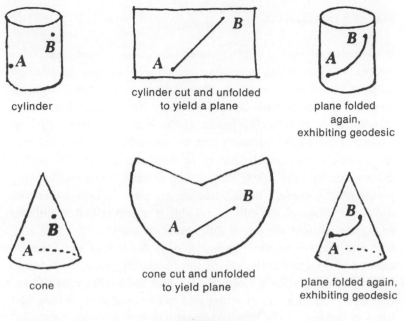

FIGURE 5

Determining a geodesic on a cylinder or cone

and a mathematician and he was coinventor with Isaac Newton of calculus. Rising to the bait, Leibniz wrote Bernoulli in July 1698 that he had thought of the problem even before seeing Bernoulli's challenge, but had not yet been able to solve it. However, he did have this idea: suppose that the surface on which we want to draw a geodesic is thought of as composed of many tiny bits of planes. If R and S are two nearby points on the surface, then in order to draw the geodesic between R and S, we allow the planes on which R and S are located to intersect. On their line of intersection, we locate a point T such that the sum of the distances from T to R and S is a minimum. Note what Leibniz is attempting to do here. He would like to use the spider-and-fly principle on a certain developable surface that approximates a very small portion of the true surface.

John Bernoulli answered Leibniz somewhat patronizingly that though Leibniz' idea was a good one, and Bernoulli had also worked with it, it did not lead him to a proper goal. But he admitted proudly to Leibniz that he had found the secret of geodesics.

It consisted of the fact that *"planum transiens per tria quaelibet puncta proxima linaea quaesitae debeat esse rectum ad planum tangens superficiem curvam in aliquo istorum punctorum."* In plain English, Bernoulli had discovered that "the plane that passes through three neighboring points of the geodesic curve, must be perpendicular to the tangent plane to the surface at these points." Bernoulli's Latin may be easier to understand than his mathematics, for by "the plane that passes through three neighboring points of the curve," he is referring to what is now called the *osculating plane* of the curve, and this is a notion that is difficult to explain without the use of differential calculus.

But there is another formulation of the law of the geodesic curve that does not make use of the osculating plane. In a sense, it lies halfway between Leibniz' idea and Bernoulli's idea. These two men lived in a day of string and spiders and flies. If they had lived in the day of the Band-Aid and had had occasion to spread one over a skinned knuckle, the formulation would not have escaped them.

Let's begin by noticing two things:

1. If two surfaces are tangent along a curve, and the curve happens to be a geodesic on one surface, it will also be a geodesic on the second surface. This can be made plausible by imagining an infinitely thin string lying between the two surfaces and stretched between *A* and *B*. If it is stretched taut for one surface, it will be taut for the other as well. Hence, if it lies along a geodesic on one, it will lie along a geodesic on the other.

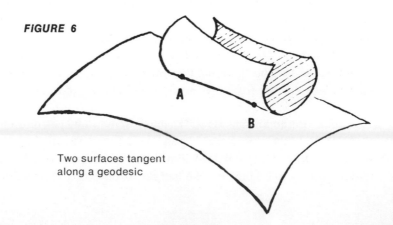

FIGURE 6

A

B

Two surfaces tangent
along a geodesic

2. If a surface has a short geodesic drawn on it, then along this curve we can apply a thin strip of flexible paper so that the paper is tangent to the surface along the curve. This can also be made plausible: a geodesic comes about by stretching a string on a surface, but if the string is "widened," it becomes a strip, and the strip will be tangent to the surface.

Putting these two observations together, we obtain what may be called the Band-Aid principle for geodesics.

Take a piece of paper and draw a straight line on it. Along a sufficiently small portion of a geodesic, we can glue a portion of this paper in such a manner that the piece of paper is tangent to the surface along the geodesic, and the geodesic coincides with the deformed straight line lying on the paper.

In order to get a feeling for what this theorem says, try this experiment. Take a strip of paper that has a straight edge. Now try to apply this edge to, say, the 60-degree parallel of latitude on a globe. It can be done, but notice how. you must lift the paper up at an angle to the surface of the globe. It cannot be done with the paper tangent to the surface.

surface with geodesic arc drawn

Band-Aid applied to geodesic arc

FIGURE 7
Illustrating the Band-Aid principle

Leibniz simply did not go far enough. Instead of applying planes to the curve at nearby points, he should have noticed that he could apply a developable surface to whole arcs of the geodesic.

What does higher mathematics make of all this? Quite a bit. Bernoulli's principle (or the Band-Aid principle) leads to the exact mathematical description of the differential equation of geodesics on a surface. Bernoulli did not prove his principle. A quarter of a century later, in 1728, Leonhard Euler gave a proof that was based on certain ideas of mechanics. It was not until 1806 that a first completely mathematical proof was given by Joseph Louis Lagrange.

Geodesics are of great importance in pure mathematics, in cartography, and in mathematical physics. One of the high points of their career came in the middle of the nineteenth century when William Rowan Hamilton formulated a principle which can be interpreted as saying that the universe goes through its gyrations along a geodesic of a certain sort. Although this is a direct descendant of stretching strings on globes, it requires a great deal of knowledge to be made comprehensible. I mention it here only to whet the appetite.

16

THE SPIDER AND THE FLY

THE NOTION OF A GEODESIC was introduced in "The Band-Aid Principle." Here, we offer a slight variation to show some repercussions of the theme in the plane. Recall that the problem of the spider and the fly originated from a situation that is illustrated in Figure 1. The hungry spider, located on the west wall at

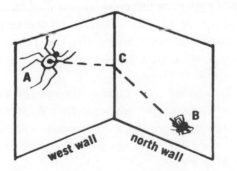

FIGURE 1

A, wants to find the shortest path along the walls to the juicy fly on the north wall at *B*.

We considered various paths from *A* to *B* through a point *C* on the northwest corner of the room, tacitly and intuitively eliminating from consideration any possible path via the ceiling or the floor. This reduced the problem to finding the geodesic from *A* to *C* and from *C* to *B*. Knowing that the straight line segment from *A* to *C* has the shortest length of all curves between A and C, and likewise for the segment from C to B, we turned our attention next to the choice of the point C so that the total length of the two segments is minimum. This was simply solved by unfolding the two walls and flattening them out (developing) so that they lie in one plane as shown in Figure 2. It was then easy to see where to place the point along the northwest corner: precisely on the segment, $\overline{AB}$.

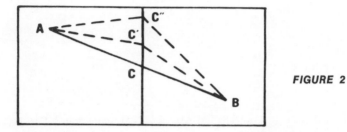

FIGURE 2

We are now ready to tackle a more complicated problem. The story is the same but the scene has changed—the spider and the fly are now at opposite ends of the room. Suppose the room is 40 feet long, 20 feet wide, and 20 feet high. The spider (A) is located on the west wall midway between the north and south boundaries and 1 foot from the floor; the fly (B) is located on the east wall midway between the north and south boundaries and 1 foot from the ceiling, as in Figure 3. Again, the problem is to find the shortest path.

As anyone who has opened out a cracker box can tell us, this rectangular prism is a "developable" surface—that is, it can be "opened" out and flattened, as we did with the two walls in the previous problem. But since our room has six sides, there is more than one way of flattening it out.

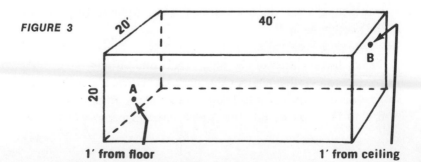

FIGURE 3

20'   40'

20'

A

B

1′ from floor   **1′ from ceiling**

a. end walls hinged to floor

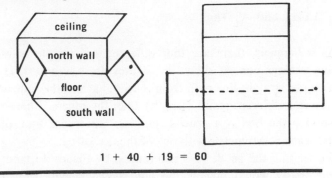

$$1 + 40 + 19 = 60$$

b. end walls hinged to north wall

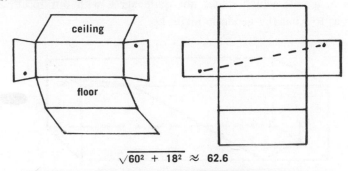

$$\sqrt{60^2 + 18^2} \approx 62.6$$

c. west wall hinged to floor; east wall hinged to south wall

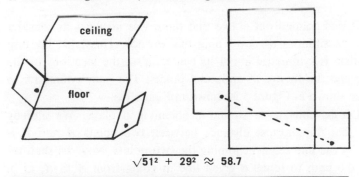

$$\sqrt{51^2 + 29^2} \approx 58.7$$

d. west wall hinged to floor; east wall hinged to ceiling

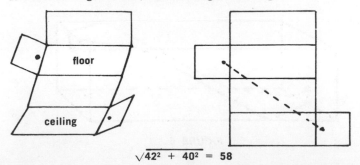

$$\sqrt{42^2 + 40^2} = 58$$

FIGURE 4

As it happens, there are four different and unique ways of flattening out our box, as shown in Figures 4a, b, c and d. If we now draw our straight-line path from *A* to *B,* each of the four ways shown gives a different result (length) than the others. Therefore, *how* we skin the box is a crucial question (there are four other ways, but each is symmetric with one of those shown).

We see that the geodesic is determined by Figure 4d, and the path the spider would take is the one shown heavily drawn in Figure 5, a path which might not quite agree with our intuitive notion as to what the geodesic might be.

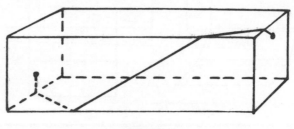

FIGURE 5

It was pointed out above that there was no need to consider *all* the possibilities of developing this surface. This is so because this prism is symmetric about its center; also, the locations of the spider and the fly are at antipodal points. The path symmetric to the one shown in Figure 5 is shown in Figure 6.

Let us consider a similar problem in a plane. We already know that the shortest distance between two points *A* and *B* is given by the line segment joining the two points. Now we shall require the path to touch a given line in going from *A* to *B,* as in Figure 7. This means finding a point *C* on the line so that the union

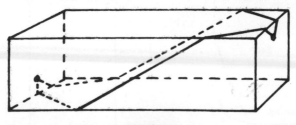

FIGURE 6

of $\overline{AC}$ and $\overline{CB}$ has the shortest length possible; thus, it is a geodesic problem. There are various ways of attacking this problem. Of these, I think the neatest one comes from an area in mathematics going by the label of geometric transformations. Here, we consider

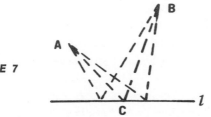

FIGURE 7

a point B' on the other side of l so that $\overline{BB'}$ is perpendicular to l and is bisected by l, as in Figure 8. B' is called "the image of B by reflection in the line." Pick any point C on l; it can be seen that $\overline{CB}$ is congruent to $\overline{CB'}$, so $AC + CB = AC + CB'$. Now it is clear which point C we must choose so as to get the shortest length required. C is the intersection of segment $\overline{AB'}$ with line l.

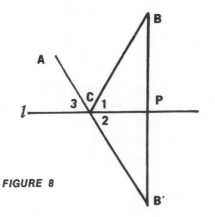

FIGURE 8

All this doesn't seem to have much to do with billiards, except possibly that our spider friend was brought up in a billiard room. In billiards, it is known that when a ball hits the side of the table, angle 3 will be congruent to angle 1; so our geodesic problem turns out to be equivalent to the problem of aiming cue ball *A* so that it would hit *B* after "cushioning" off from the side of the table. (Bouncing rays off flat mirrors may also be treated in this manner.)

We can easily show that our geodesic construction fulfills this requirement. Let *P* be the intersection of $\overline{BB'}$ with *l;* then it can be proved that $\triangle BCP$ is congruent to $\triangle B'CP$ and hence angle 1 ≅ angle 2. Also, since $\overline{AB'}$ is a straight line segment, angle 2 and angle 3 are vertical angles and so now angle 3 ≅ angle 2 ≅ angle 1.

What if we were required to hit *B* with cue ball *A* after cushioning off from two sides *l* and *l'*, as in Figure 9?

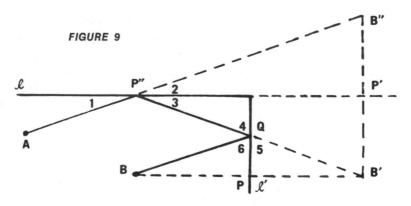

FIGURE 9

To do this, we first get *B'*, the image of *B* by reflection in line *l';* then get *B''*, the image of *B'* by reflection in line *l*. By the congruences of $\triangle B''P'P''$ with $\triangle B'P'P''$, and of $\triangle B'PQ$ with $\triangle BPQ$, we can see that aiming cue ball *A* at *B''* will send the ball to *B* after two cushions. Clearly, this method can be extended to take on any number of additional cushions—and this knowledge should enable me to earn a ranking comparable to that of the late great Willie Hoppe—except that I haven't yet learned to hold a cue properly.

17

A WALK IN THE NEIGHBORHOOD

AFTER HANGING up a picture one day, I stepped back clear across the room to see whether it hung straight or not. I had no sooner reached the other side when the shade of Aristotle tapped me on the shoulder and asked, "How could you possibly do that?" I tried to explain that I wanted to see how the picture hangs, but the shade interrupted, "That is not the point at all; hasn't it been demonstrated that you cannot walk across the room?" Only then did I realize that the shade of Aristotle had in mind one of the arguments of Zeno of Elea, a Greek philosopher who lived about 485 B.C.

Zeno had proposed four paradoxes relating to space, time and motion. The first of the paradoxes proposes to show that motion does not exist, and demonstrates this as follows:

1. To walk across the room, you must first walk half of the distance; this requires a finite amount of time.

2. To cover the remaining distance, you must first cover half of its distance; this requires more time.

3. Whatever the remaining distance, it may be halved and the first half must be traversed before reaching any portion of the second half; each half-distance requires a finite amount of time.

4. There is an infinite number of half-distances, requiring therefore an infinite amount of time to complete the journey.

This argument of Zeno is referred to as the "Dichotomy"

(dichotomy means "a cutting in two"). Aristotle explained that we have a paradox here, because while we were willing to divide a line segment into infinitely many parts, we were reluctant to divide time into infinitely many parts.

Let us first assume that we don't change our speed as we walk across the room. If it takes half a minute to cover half the distance across the room, then it would take one fourth of a minute to cover one fourth the distance, etc. Let us now look at the (cumulative) number of minutes we would need as we progress across the room:

$$\frac{1}{2}, \quad \frac{1}{2}+\frac{1}{4}, \quad \frac{1}{2}+\frac{1}{4}+\frac{1}{8}, \quad \frac{1}{2}+\frac{1}{4}+\frac{1}{8}+\frac{1}{16}, \cdots$$

or

$$\frac{1}{2}, \quad \frac{3}{4}, \quad \frac{7}{8}, \quad \frac{15}{16}, \cdots$$

This gives us a sequence which we can easily describe if we note that the denominator of each term is a power of 2 and that the numerator is one less than the denominator; that is, the nth term is $\frac{2^n - 1}{2^n}$.

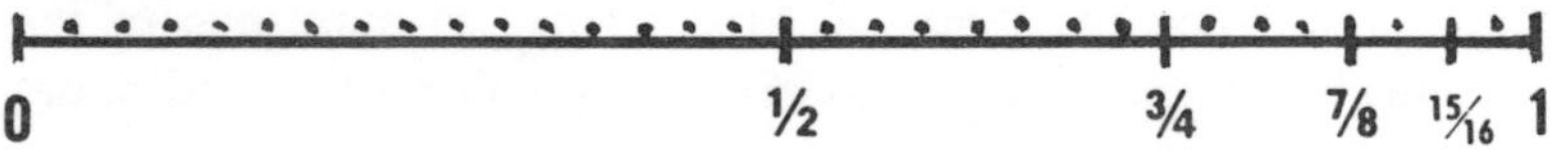

FIGURE 1

From the sequence and from the graph in Figure 1 we get an intuitive idea that the total time gets closer and closer to one minute but does not exceed one minute. We say that 1 is a limit point of this set of numbers.

To describe mathematically what we mean by a limit point, we must first explain what we mean by a *neighborhood* of a point. A neighborhood of a point is the set of all points within a certain distance from the point. More precisely, a neighborhood of a point is the set of all points whose distance from the point is less than a certain radius r. For points on a line, if p is the coordinate of the point, then the r neighborhood of the point is the interval from

$p - r$ to $p + r$, excluding the endpoints. To find a neighborhood of a point in the line is easy; just set the compass at the point, and with r as radius, strike arcs intersecting the line (*see Figure 2*). Then all the points within the interval are in the neighborhood. In particular, p is in the neighborhood. To find a neighborhood of a point in the plane is not much more troublesome; again set the compass at the point and draw a circle with r as radius (*see Figure 2*). Then all points in the interior of the circle are in the r neighborhood of the point; the circle is the boundary but is not itself in the neighborhood. Similarly, in 3-space, the neighborhood is the interior of a sphere, and in n-space it is the interior of an n-dimensional sphere. This is why a neighborhood is sometimes referred to as a "spherical" neighborhood. It is entirely possible to define a neighborhood of a point without the requirement that the point is to be centrally located, and this approach is indeed sometimes adopted. A neighborhood of a point is then simply defined as the interior of a sphere (circular regions and line segments are included as 2- and 1-dimensional spherical regions) containing the point. A neighborhood so defined is said to be *asymmetric,* and theorems developed on this basis are entirely equivalent to corresponding theorems on the basis of symmetric neighborhoods. That such equivalency is possible can be seen when it is recog-

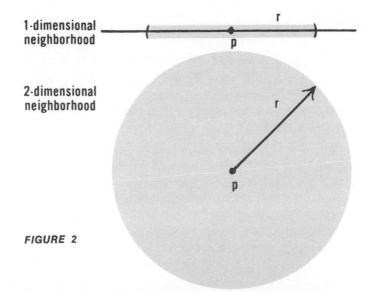

1-dimensional neighborhood

2-dimensional neighborhood

FIGURE 2

nized that if there is a symmetric neighborhood of a point, then an asymmetric neighborhood can be found, and conversely, if an asymmetric neighborhood exists, a symmetric neighborhood likewise exists.

With either interpretation, the concept of a limit point can be described now in terms of neighborhoods. Given a set of points, for example,

$$\frac{1}{2}, \frac{3}{4}, \frac{7}{8}, \frac{15}{16}, \cdots$$

then a point p is a *limit point* of the set if every neighborhood of p contains a point of the set other than p. Note that a limit point of the set is not itself necessarily a point of the set. The crucial point is that no matter how small a radius is taken, if p is to be a limit point, then each neighborhood of p must contain a point of the set other than p; that if *any* neighborhood of p can be found which does not contain a point of the set then p is not a limit point of the set.

With the above definition, we can show that 1 is a limit point of the above sequence. To do this, take any radius r. The requirement is then to find a number in the set such that

$$1 - \frac{2^n - 1}{2^n} < r.$$

Since $\dfrac{2^n - 1}{2^n} = 1 - \dfrac{1}{2^n}$, the requirement may be restated as $\dfrac{1}{2^n} < r$. Since $\dfrac{1}{2}, \dfrac{1}{4}, \dfrac{1}{8}, \cdots$ are increasingly small, for any given radius r, we can always find a large enough n such that $\dfrac{1}{2^n} < r$; for any n that meets this requirement, $\dfrac{2^n - 1}{2^n}$ is in the r-neighborhood of 1.

We can also show that no other point is a limit point of this set. If q is such a point and if $q > 1$, then a neighborhood of radius $q - 1$ contains no point of the set; so q is not a limit point. If $q < 1$, then q is between two points, q' and q'', of the set, where

$$q' = \frac{2^n - 1}{2^n} \quad \text{and} \quad q'' = \frac{2^{n+1} - 1}{2^{n+1}}.$$

If the radius of the neighborhood of q is the smaller of the distances from q to q' or to q'', then this neighborhood contains no point of the set; so here again q is not a limit point. Geometrically, we can visualize the situation thus. The sequence of points on a line progressively approaches the point with coordinate, 1. If q is less than 1, then it is located between two successive points, q' and q'', of the sequence, and it is a finite distance from either q' or q''. Take the smaller of these distances; then this neighborhood of q contains no point of the sequence, and q is not a limit point.

Finding the limit point may or may not be an easy thing to do. In the case of Zeno's dichotomy it was quite simple. This was an example of a geometric series where each term is a constant multiple of the term before it. If S_n is the sum of the first n terms, we have

$$S_n = \frac{1}{2} + \frac{1}{4} + \frac{1}{8} + \ldots + \frac{1}{2^n}.$$

Multiplying S_n by 2 and subtracting S_n from the product, we get

$$2\,S_n = 1 + \frac{1}{2} + \frac{1}{4} + \ldots + \frac{1}{2^{n-1}}$$

$$S_n = \quad\ \ \frac{1}{2} + \frac{1}{4} + \ldots + \frac{1}{2^{n-1}} + \frac{1}{2^n}$$

$$S_n = 1 - \frac{1}{2^n}$$

so that $S_n = \dfrac{2^n - 1}{2^n}$. This agrees with the expression we obtained by inspection for the nth term of the sequence.

The limit point may be found by noting that each of the two series in the above subtraction goes on and on, and that for every term in the bottom series there is a matching term in the top so that the subtraction results in 1. The limit points of the following sequences are also easily obtained by inspection and proved to be such by the definition for limit point:

$$1.9, \quad 1.99, \quad 1.999, \quad 1.9999, \ldots$$
$$2.1, \quad 2.01, \quad 2.001, \quad 2.0001, \ldots$$

The limit point of each is 2. The nth term of the first sequence is a variation of the geometric series in this form:

$$1 + \frac{9}{10} + \frac{9}{100} + \frac{9}{1000} + \cdots,$$

and the nth term of the second may be expressed as

$$3 - \frac{9}{10} - \frac{9}{100} - \frac{9}{1000} - \cdots$$

Sometimes it may not be easy to find the expression for the nth term; even if we have the nth term, it may not be easy to find the limit point. Then again, the form for the expression may look more forbidding than it is in reality. As an example of a case where the bark is worse than the bite, take a look at the following:

$$\sqrt{2}, \quad \sqrt{2 + \sqrt{2}}, \quad \sqrt{2 + \sqrt{2 + \sqrt{2}}}, \ldots$$

We may first express each of the above terms in its decimal approximation

$$1.414, \quad 1.847, \quad 1.962, \quad 1.990, \quad \ldots,$$

from which we can guess that the limit point is 2. If we do not have the patience to attempt this brute-force method, we may instead try to find directly the positive number x such that

$$x = \sqrt{2 + \sqrt{2 + \sqrt{2 + \cdots}}}.$$

Bracing ourselves for the task, we square both sides to get

$$x^2 = 2 + \sqrt{2 + \sqrt{2 + \sqrt{2 + \cdots}}}.$$

This doesn't look much better until we observe that the right-hand radical is exactly what we had labeled as x in the first equation. So the equation reduces to $x^2 = 2 + x$, or equivalently, $x^2 - x - 2 = 0$. From this, we get 2 as the positive solution.

It can be seen that the same technique will work for

$$x = \sqrt{c + \sqrt{c + \sqrt{c + \cdots}}},$$

which reduces to a simple quadratic equation $x^2 - x - c = 0$ whose positive solution is given by $x = \dfrac{1 + \sqrt{4c + 1}}{2}$. For $c = 3$, the solution is $\dfrac{1 + \sqrt{13}}{2}$ and for $c = 4$ it is $\dfrac{1 + \sqrt{17}}{2}$.

We may next ask when we can expect positive integers for solution when c is an integer. For this to happen, the discriminant $4c + 1$ must be a perfect square. We can start by setting up a table for c and for the corresponding discriminant d and see when we do get a perfect square for d:

c	0	1	2	3	4	5	**6**	7	8	9	10	11	**12**	. . .
d	**1**	5	**9**	13	17	21	**25**	29	33	37	41	45	**49**	. . .

Picking out the ones that are perfect squares for d in this table,

c	0	2	6	12	20	30	42	56	. . .
d	1	9	25	49	81	121	169	225	. . .

If we go far enough in the table and if we are pretty sharp with our hunches, we may soon discover that the sequence for the c's is of the form: $1 \times 0, 2 \times 1, 3 \times 2, 4 \times 3, \ldots, n(n - 1)$. Since we cannot always depend on our antenna beaming satisfactorily, we require directly that x is to be an integer in the equation

$$x = \frac{1 + \sqrt{4c + 1}}{2} \text{ or equivalently, } \sqrt{4c + 1} = 2x - 1.$$

Solving for c, we get $c = x(x - 1)$, where x is an integer; this tells us what we wanted to know without guesswork. Moreover, this tells us that any integer larger than 1 may be expressed in the form

$$\sqrt{c + \sqrt{c + \sqrt{c + \ldots}}}.$$

For example, this expression for $x = 4$ is obtained by using $c = 4 \cdot 3$, for $x = 5$ by using $c = 5 \cdot 4$, and in general for $x = n$ by using $c = n(n - 1)$.

Using a little hindsight, we could have found this from

$$x^2 - x - c = 0$$

by asking what factors of c will this be true of. The answer, of course, is when the difference of the factors is the coefficient of the x term, which is -1. So this is true when $n - 1$ and n are factors of c.

By way of rounding things up, I should really tell what eventually happened to the shade of Aristotle. I finally got up enough nerve to ask, "By the same token that I cannot walk across the room, I cannot walk across half the room; nor can I walk across half of that; nor half of that. If I cannot get from there to here, neither can you; how did *you* ever get from there to here?" Sheepishly, the shade said, "I had neglected to consider that," and with that, he gracefully faded away into the past.

18

DIVISION IN THE CELLAR

I LIKE TO DO WOODWORKING in my spare time, and the month hardly passes in which I do not go down to my workbench in the cellar and putter around. The workbench is not terribly well equipped with tools, but worse than that, the tools I have are generally scattered through the house.

Here is a typical situation. I'm in the cellar banging away at a project. My pockets are bulging with tools. I have a pencil behind one ear, a crayon behind the other, and a mouthful of tacks and nails. In the course of this furious activity, I find that I must divide a plank into a number of parts of equal length. I look for a ruler or a yardstick and I can't find one. I look for a tape measure or a piece of string I can fold up. Nothing doing. But I remain calm. I've met this crisis before, and I'll solve the problem by using a method of successive approximations.

Let's suppose I have a plank I'd like to divide into four equal parts. Here is what I do. I take another plank whose length I judge to be greater than a quarter of the plank I want to divide. On this auxiliary plank, I mark off a trial length which I will use as a first approximation to the answer. I lay this trial length on the plank four times. If my guess was good, laying off the length four times will just fill up the plank. But this rarely happens. If I have underestimated the quarter, there will be a certain residual left over on the plank. If I have overestimated the quarter, there will be a residual left over on the auxiliary stick (*see drawing*).

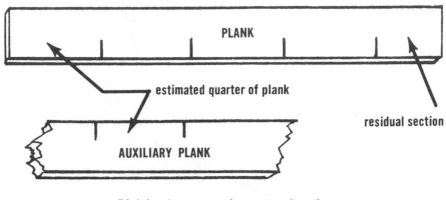

Division by successive approximation

Now I work with the residual in the same manner. I estimate one quarter of the residual, and add it to or subtract it from my original guess, according to whether I underestimated or overestimated in the first place. It will often happen that I have not guessed one quarter of the residual with sufficient accuracy, and so, just as big fleas have little fleas upon their backs to bite 'em, the residual will itself have a residual, still smaller. I continue in this way, estimating quarters of residuals, and eventually, I have divided my plank.

This, then, is what I do when I can't find a ruler.

Now it might strike you that this is a very primitive way of dividing a length. And so it is. And yet, the method of successive approximations forms one of the principal methods by which computation is performed on electronic computers. A guess is made at the answer, one sees how ·closely the answer solves the problem, and then one corrects the answer on the basis of this information. Shoot, miss, and correct; shoot, miss, and correct, over and over again, perhaps hundreds of times, until the answer has been obtained to within the accuracy desired.

A simple instance of successive approximations used on a high-speed computer is for the computation of square roots. Here is how it works. Suppose we would like to compute $\sqrt{7}$ to 4 decimal places. Now $\sqrt{7}$ is that number which when multiplied by itself yields 7. An alternative way of putting it is this: $\sqrt{7}$ is that number which when divided into 7 yields $\sqrt{7}$ over again. In symbols, if $x = \sqrt{7}$, then $x^2 = 7$ and $x = \dfrac{7}{x}$. Now take a guess at the value of $\sqrt{7}$. Say 2. Divide 2 into 7: $\dfrac{7}{2} = 3\frac{1}{2}$. If 2 had been $\sqrt{7}$, we would have gotten 2 back again. If we had guessed $2\frac{1}{2}$ to start with, then $\dfrac{7}{2\frac{1}{2}} = 2\frac{4}{5}$. This would be a better guess, for the divisor and the quotient are closer together.

Our job is to find the value where the divisor equals the quotient. Let us therefore adopt this policy, which seems altogether reasonable: Go halfway between the divisor and the quotient and use this number as a new divisor. Do this over and over again. Here is how it works out starting from a guess of 2.

NUMBER OF APPROXIMATION	"SHOOT" (DIVISOR)	"MISS" (QUOTIENT)	CORRECTION (AVERAGE OF DIVISOR AND QUOTIENT)
1	2	3.5	2.75
2	2.75	2.55	2.6477
3	2.6477	2.6438	2.6459
4	2.6459	2.6456	2.64575
5	2.64575	2.64575	•
•	•	•	•
•	•	•	•
•	•	•	•
n	x_n	$7/x_n$	$\frac{1}{2}(x_n + 7/x_n)$

Table I

Notice the way the divisors and the quotients become closer to one another, and after five lines, agree with one another to 5 decimal places. The whole process can be expressed neatly by a formula. If x_n designates the nth divisor, then the $n + 1^{st}$ divisor is given by

$$x_{n+1} = \tfrac{1}{2} \left(x_n + \frac{7}{x_n} \right)$$

This scheme is not terribly tempting for paper-and-pencil work because of the unpleasantness of division, but on an electronic computer it is rapid and quite painless.

The curious thing about the scheme is that it was first propounded more than two thousand years ago by the Greek scientist Heron, who lived in Alexandria, and was developed by Isaac Newton in the seventeenth century; but it took the facilities of modern electronic equipment to bring it into its own.

The science of obtaining numerical answers to mathematical problems is known as *numerical analysis.* It is part of numerical analysis to derive methods of successive approximation, or—as they are frequently called—*iterative methods,* and to show that when carried out over and over again, answers are obtained that can be made as close to the true answer as is desired. This is known as the *convergence* of the repeated procedure (in the previous problem, convergence is illustrated by the values of "shoot" and "miss," which come closer and closer to each other at each step).

Finding a convergent method can be a simple task or an exceedingly complex one, depending upon the nature of the problem.

There are many types of repetitive corrections which may seem natural enough, but which do not lead to a convergent process. For example, suppose that we are asked to solve the problem $x = 2x - 1$. The answer, as you can easily see, is $x = 1$, and it might seem silly to try to solve it by successive approximations. Nevertheless, let's try anyway, using the same method as for the square root, and see what happens. The formula is now

$$x_{n+1} = \tfrac{1}{2}(x_n + 2x_n - 1).$$

Taking a starting value of 2:

NUMBER OF APPROXIMATION	"SHOOT"	"MISS"	CORRECTION
1	2	3	2.5
2	2.5	4	3.25
3	3.25	5.5	4.375
4	4.375	7.75	6.0625
•	•	•	•
•	•	•	•
•	•	•	•
n	x_n	$2x_n-1$	$\frac{1}{2}(x_n + 2x_n-1)$

Table II

Notice that the shoots and the misses appear to be moving away from one another and from the answer $x = 1$. The process is not a convergent one! The last word has yet to be said by mathematicians on questions relating to successive approximations.

But let's return to the cellar and to the division of the plank that started this discussion. My experience has been that I can carry out such a division with carpentry accuracy in about four or five tries. But I would like to put this on a firm mathematical basis. I would like to show that making some reasonable assumptions as to how accurate my guesses are, the process is a convergent one. I will assume that whenever I divide any length at all into four equal parts by eye, the error I make does not exceed half the original length. This assumption is reasonable—in fact, quite conservative. Let the plank have unit length, and let d_1 be the estimated quarter and r_1 be the first residual. Then

$$4d_1 + r_1 = 1$$

where, by our assumption,

$$-\frac{1}{2} \leq r_1 \leq \frac{1}{2}.$$

Let d_2 be the estimated quarter of the first residual, and r_2 the second residual. Then,

$$4d_2 + r_2 = r_1.$$

Since r_1 is between $-\frac{1}{2}$ and $\frac{1}{2}$, and since our error in the second division can be no worse than $\frac{1}{2}$ the length of r_1, it must follow that $-\frac{1}{4} \leq r_2 \leq \frac{1}{4}$. It can be shown similarly that

$$4d_3 + r_3 = r_2 \qquad \text{where} \qquad -\frac{1}{8} \leq r_3 \leq \frac{1}{8}$$

. . . .

$$4d_n + r_n = r_{n-1} \quad \text{where} \qquad -\frac{1}{2^n} \leq r_n \leq \frac{1}{2^n}$$

If we add these equalities we obtain

$$4(d_1 + d_2 \ldots + d_n) + r_1 + r_2 \ldots + r_n = \\ 1 + r_1 + r_2 \ldots + r_{n-1}.$$

Hence,

$$4(d_1 + d_2 \ldots + d_n) = 1 - r_n$$

and

$$d_1 + d_2 \ldots + d_n = \frac{1}{4} - \frac{r_n}{4}$$

This equation tells us that the adjusted guess at the nth stage, $d_1 + d_2 \ldots + d_n$, differs from $\frac{1}{4}$ (the required length) by the quantity $\frac{r_n}{4}$. We know that this is no larger than $\frac{1}{4} \left(\frac{1}{2^n} \right) = \frac{1}{2^{n+2}}$ in size. As n becomes larger and larger, this error decreases in size very rapidly. When $n = 5$, the error is at most $\frac{1}{2^7} = \frac{1}{128}$ which is less than 1 per cent. That's pretty good shooting!

19

THE ART OF SQUEEZING

I HAVE NEVER PLAYED GOLF and I doubt that I can ever hit that little ball, but this doesn't keep me from having a theory on how I would play if I ever found myself with a club in my hands. I would give the club a mighty swing. Knowing my background on this sport, you can be quite sure that I would miss on the first whack. This is where my theory comes to the fore! When I take the first swing, *my attention will not be on how I am going to hit the ball, but on how I am going to miss it*. (I already take it for granted that I shall miss.)

The fact that I will miss the ball should not bother me; golfing is only a game. Now, if my club misses the ball by three inches to the left, then I shall do as anyone else might do—compensate by aiming more to the right. Knowing my aptitude again, I shall most likely compensate too much, with the net result of say, being two inches to the right on the second swing. So then I try to correct my correction. Maybe this time I shall be only one and one-half inches to the left. Theoretically, I can continue this process and eventually come in contact with the ball. (It is clear that my low-level golf ambition is confined to merely connecting with the ball; how it behaves after it is hit is not my concern.) If this aiming procedure sounds foolish, recall the days before radar and automation in the firing of missiles. For example, before blasting away at the Pirates of Penzance, the boys on H.M.S. *Pinafore* should really take into account the roll of the ship, its pitch, the wind velocity, et cetera. As these considerations are somewhat overwhelming, the boys do the next best thing: fire one volley, estimate the miss, fire again—much as I fire away with the golf club.

At the beginning of World War II this kind of problem received a good deal of attention; in time of war gunners cannot always depend on having the opportunity of firing another round. In the attempt to increase efficiency in scoring, it was felt that a "reasonable" approach involved solving some twenty-odd simultaneous equations in as many variables. At that time, fast and easy procedures for breaking down such problems were not generally available. One might start by displaying the twenty-odd equations in the twenty-odd variables and methodically "eliminate" variables one by one; or one might even try to work with the determinants associated with the system of equations. Unfortunately, air and naval gunners cannot afford the time required by these approaches. Hence, the hit-and-miss method was justified—especially when an experienced crew can get to the target in short order.

As an overly simplified illustration, consider target practice with the rifle. If the target is far enough, evidently aiming directly will not do, since the effect of gravity on the bullet cannot be entirely neglected. Hence the practical rule of thumb: Aim a little higher (*see Figures 1, 2*). The question as to how much higher to aim takes us right back to the hit-and-miss method; either that, or find the equation of the curve that passes through the bull's-eye of the target. In physics one learns that the parabola can be used for this purpose if the only factor to consider is the effect of gravity, neglecting other effects such as wind velocity, how much energy the bullet has spent at each stage, et cetera. This is because, if an object were dropped, the distance in feet fallen (the vertical distance) after t seconds is approximately $16t^2$; that is, $y = -16t^2$. On the other hand, if the bullet is shot out at a constant velocity and maintains this velocity v, then the horizontal distance is $x = vt$. Eliminating t from these two equations we get the equation of the path: at each instant, $x^2 = -\dfrac{v^2}{16} y$. If the velocity of the bullet is

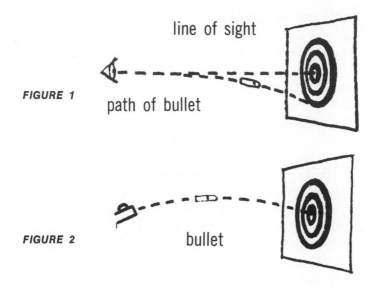

line of sight

FIGURE 1 path of bullet

FIGURE 2 bullet

4,000 feet per second, then $x^2 = -10^6 y$, which can be recognized as an equation of a parabola.

From the equivalent form, $y = -\dfrac{x^2}{1,000,000}$ we can see that for small values of x, our deviation from the bull's-eye is not enough to worry us. For example, standing one foot from the target, $x = 1$ and $y = -10^{-6}$; in other words, the bullet hits one millionth of a foot below the bull's-eye. We see from the same equation, that if x is 1,000 then $y = -1$; that is, standing 1,000 feet from the target (which means a little less than two blocks away), we can expect the bullet to be one foot from dead center—enough to disqualify the claim that we have found the mark. Even if the rifle were tipped so that it is not aimed along the horizontal axis, the resulting equation turns out to be a parabola—provided that we do not consider other factors that may influence the path of the bullet.

Assuming now that we have found an equation that will lead us to a satisfactorily close estimate of the path, we must next tackle the equation; this in itself may be something of a problem. If we come up with a second-degree equation, we know that the solution can always be expressed, at most, with radicals; our quadratic formula gives the possible solutions immediately. With a third-degree equation (a cubic), there are ways of transforming the equation in terms of another variable so that the resulting equation can be solved by a formula involving taking cube roots. Similarly, fourth-degree equations may be broken down so that simpler methods may be applied. In the case of equations of fifth degree or higher, the story may be quite different; some of these equations may be broken down fairly easily; but again, there are equations that cannot be solved by radicals. The first person to demonstrate that a fifth-degree equation (a quintic) is not always solvable by such means was the young Norwegian mathematician Niels Henrik Abel (1802–1829). Before then, it was usually assumed that this is always possible; merely that the methods had not been found for some of the sticklers.

Since we may not always be able to find an exact solution for a quintic, we may have to be content with an approximate solution. Let's see how a typical case may be handled. Consider the equation $y = x^5 + 2x^4 + 3x^3 + 4x^2 + 5x + 6$. We note that if x is 0 or greater, then y is positive: in particular, if x is 1, then y is 21. Also, if x is -2, then y is -12. Plotting these two points on the (x,y) plane, we have the situation illustrated in Figure 3. It can be proved that the graph of this equation is a continuous curve, so we can expect that somewhere between these two points, the graph crosses

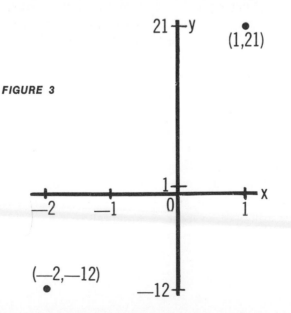

FIGURE 3

the horizontal axis; that is, somewhere between $x = -2$ and $x = 1$, there is at least one x such that y is 0. Thus, this is analogous to my golf game; if I aim at $x = 1$, the y is 21 inches too high; if I aim at $x = -2$, the y is 12 inches too low. Trying $x = 0.5$, we get $y = 10$, and in this way, we wobble back and forth getting a table of values (some approximate) as the following:

x	1	-2	0.5	-1	-1.4	-1.5	-1.45
y	21	-12	10	3	0.9	-0.6	0.51

(These are the points labeled a, b, . . . , g in Figure 4.)

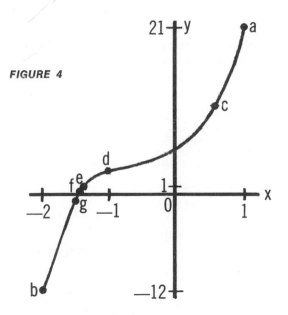

FIGURE 4

Examining what we have just done in the above example, we may see that we are zooming in on the point whose y value is zero. From the first two tries, we decided that the point must be between a and b, since y is positive for one point and negative for the other; so x must be between 1 and -2. Next, we tried $x = 0.5$ and found that the corresponding y is 10. This indicates that the point we want must be between b and d (x must be between -2 and -1), and so forth. Using the abbreviation $[m,n]$ to mean that x is in the interval from m to n $(m \le x \le n)$, we find that we were getting closer to the x value for which y is zero by the following sequence of intervals:

$$[-2, 1], [-2, 0.5], [-2, -1], [-2, -1.4], [-2, -1.5],$$
$$[-1.5, -1.4], [-1.5, -1.45].$$

Because each interval in the sequence contains the next, this is called a sequence of *nested* intervals; moreover, this is called a *contracting* sequence, because each interval gets smaller and smaller. Thus, by a contracting sequence we squeeze down on the value of x ("zero in" on the bull's-eye) until eventually we get as close to the actual value of x as we wish (*see Figure 5*). Recall that this is another way of saying that x is the limit point of the sequence: Every neighborhood of x contains a point of the set other than x (*see* "A Walk in the Neighborhood").

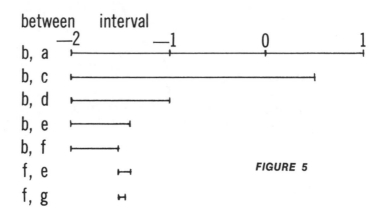

FIGURE 5

Although he may not put it in such language, this is essentially what an engineer might do in getting an approximation by graphing. The graph is first roughly sketched to get an idea of the portion of the curve he wants to look at more critically. Then comes a meticulous plotting of points in the neighborhood where his efforts are likely to be more fruitful. If necessary, successive graphs are successively magnified until a satisfactory degree of accuracy is attained. The method may be modified by bringing in the use of a computer. For example, the profile of the pressure pattern may be obtained by locating pressure sensors all over one end of the wind tunnel and analyzing their reports by a computer—followed by a closer examination of a particular critical area if necessary.

Mathematically, various techniques, such as Horner's method, Newton's method and other methods of numerical approximation, have been devised to help us decide how much correction we can apply for a better next estimate. The processes are time-consuming and call for a lot of mechanical computation. Fortunately, this is the type of activity a computer is especially suited for. Note, though, that in the situations we have described such as golfing and rifle practice, we were concerned with fixed targets. When we must deal with a moving target such as shooting for the moon, complications arise. These *pursuit problems* involve differential equations, which, fortunately again, may be handled to a large extent by computers. Even so, instructions must be programed into the computers so that the orbit may be constantly corrected during the journey. These constant corrections illustrate again how one squeezes down to a point.

As our final illustration we may observe that this technique lurks in the shadows behind our long-division process. For example, in the problem,

$$\frac{1.\,\cdot\,\cdot\,\cdot}{4.7\,\overline{)9.2563}}\;,\quad \frac{1.9\,\cdot\,\cdot\,\cdot}{4.7\,\overline{)9.2563}}\;,\quad \frac{1.96\,\cdot\,\cdot\,\cdot}{4.7\,\overline{)9.2563}}\;,\;\cdot\,\cdot\,\cdot$$

we first estimate that the quotient is in the interval [1, 2], since $1 \times 4.7 = 4.7$ and $2 \times 4.7 = 9.4$. Next, we locate the quotient in the interval [1.9, 2.0]; ($1.9 \times 4.7 = 8.93$ and $2.0 \times 4.7 = 9.40$). Next, [1.96, 1.97], etc. Thus we see that the methods may vary, but the notion is the same. In each case we squeeze down on the point of interest. Since I happen to know how to do long division, perhaps I should go out tomorrow and putter around the golf course for at least nine holes.

20

THE BUSINESS OF INEQUALITIES

"IF THEY WERE laid end to end they would stretch as far as . . ."

How often have you heard this expression, applied perhaps to all the blood vessels in a person's body, the strands of thread in a spider's web, or all the college students who sleep in class. The total distance may be so many miles, so many times around the earth, or to the moon and back, et cetera. This device of laying things end to end is very effective because it is natural to visualize things laid out in this fashion.

On the mathematical side, this technique is very helpful. For example, finding the perimeter of a polygon is essentially laying the sides of the polygon end to end on a line, and from this, arriving at a measure (*see Figure 1*).

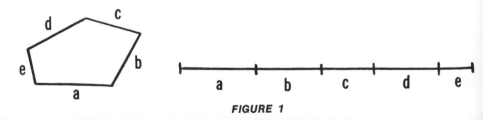

FIGURE 1

This type of thinking is also quite useful for certain exercises on deductive reasoning, as in the following problem: 25 doctors and

10 lawyers are gathered in a room. What can be deduced about the number of doctors who are also lawyers?

If all the 25 doctors may be grouped to one side of the room and all the 10 lawyers to the opposite side of the room, then no one in the room is both a doctor and a lawyer (*see Figure 2*). This separation may be illustrated equally by stretching out the doctors and the lawyers in an end to end fashion as in Figure 3.

FIGURE 2

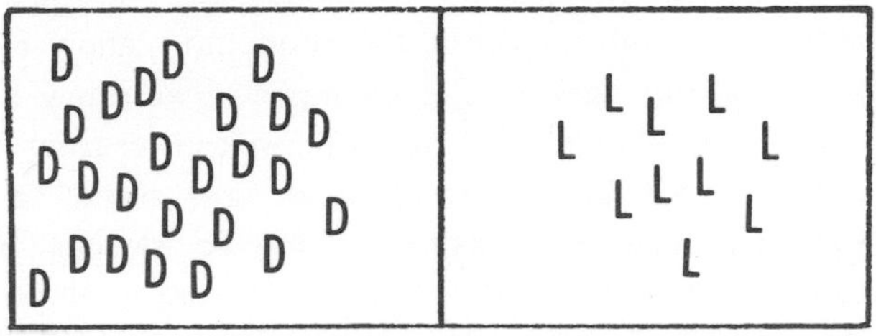

FIGURE 3

If there is to be no overlapping of the doctor and lawyer groups, we can see that there must be at least 35 persons in the room. To determine how much overlapping we may have for this example, we match a doctor with a lawyer for each individual who is both a doctor and a lawyer and consider this matched pair as one individual. The problem is, then, how many such one-to-one matchings do we have? For this purpose, we can draw a connecting curve from one D to one L for each matched pair, or picture the matching by telescoping the two lined-up groups, showing each matched pair by the arrangement ($\frac{D}{L}$). For example, Figure 4 indicates how sets of 25 doctors and 10 lawyers may be such that there will be 22 doctors who are not lawyers, 7 lawyers who are not doctors, and 3 who are both doctors and lawyers. We can see that the most these two groups can overlap is when every person in the smaller group also belongs to the larger group; in this case, when all 10 lawyers are also doctors.

doctors
doctor-lawyers
D D D D D D D D D D D D D D D D D D D D D D D D D
L L L L L L L L L L
lawyers

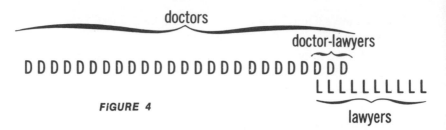

FIGURE 4

Changing the question to include some information about the total number of people in the room may allow us to be more specific in our answer. For example, if we know that there are, all together, 30 persons in the room, we can see from Figures 5a and 5b that the possible number of doctor-lawyers may be from 5 to 10; we have thus narrowed the range from the previous range of 0 to 10 doctor-lawyers. If we also know that there is no one but doctors or lawyers in the room, then we also know that there are exactly 5 doctor-lawyers.

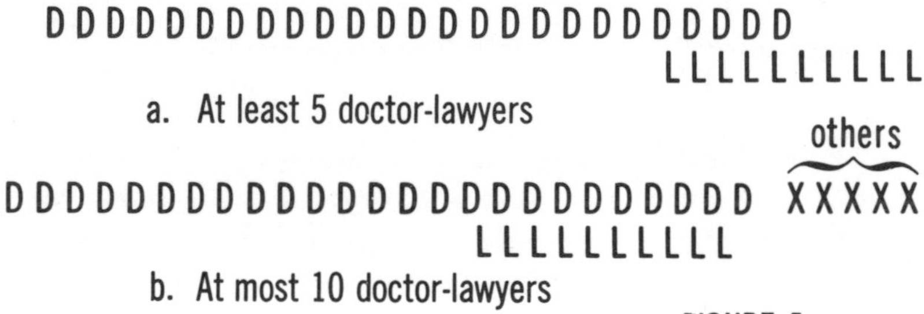

D D D D D D D D D D D D D D D D D D D D D D D D D
L L L L L L L L L L
a. At least 5 doctor-lawyers
others
D D D D D D D D D D D D D D D D D D D D D D D D D X X X X X
L L L L L L L L L L
b. At most 10 doctor-lawyers

FIGURE 5

If *A* and *B* are two sets of objects, then the *intersection* of *A* and *B* is the set of objects *A* and *B* have in common (the intersection of more than two sets is similarly defined as the set of objects all the sets have in common). Thus, the intersection of doctors and lawyers in Figure 4 are the three individuals who are members of both groups. This definition of intersection agrees completely with our notion of intersection, say, of two lines as the set of all points common to the two lines, or of two planes as the set of all points common to the two planes, and so on.

In our example it was convenient for us to indicate overlapping of the doctor and lawyer groups along chunks of a line. If we represent the set of all objects belonging to A by points in a region (labeled A in Figure 6) and B by points in another region, we can also picture the intersection of A and B as the set of points in the overlapping region (the shaded portion in the figure); this is the way we shall represent intersection of regions in the plane.

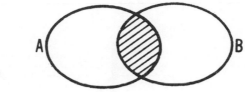

FIGURE 6

In the same sense that doctor-lawyer is a set intersection, the set of all pairs (x, y) satisfying simultaneous equations is an intersection in algebra. When we say that $(10, 2)$ is the solution for

$$\begin{cases} x + y = 12 \\ x - y = 8 \ , \end{cases}$$

for example, we mean that of all pairs (x, y) such that $x + y = 12$ and all pairs (x, y) such that $x - y = 8$, the pair that these two sets have in common is when x has the value 10 and y has the value 2; that is, $(10, 2)$. So in solving this by graphing, we find the common pair (x, y) by locating the intersection of the graphs for $x + y = 12$ and $x - y = 8$, since each graph represents the set of points for which each algebraic statement is true.

Solving simultaneous equations is quite straightforward; now what about inequalities? The problem is still to find the common set of points indicated by the algebraic statements. However, the picture has been changed considerably. Let's look at the system of inequalities, $x + y \leq 12$ and $x - y \leq 8$. Changing each of these to the equivalent statements,

$$\begin{cases} y \leq 12 - x \\ x \leq 8 \ + y, \end{cases}$$

we see that the first inequality marks all the points in the graph on or below a certain boundary indicated by $12 - x$, and the second inequality marks all the points in the graph on or to the left of the boundary indicated by $8 + y$. So the graph for the first is the (closed) half plane shown by the vertical shading on or below $y = 12 - x$ in Figure 7, and the second is shown by the horizontal shading on or to the left of the line $x = 8 + y$. The intersection is then represented by the wedge that has been doubly shaded and every point in this wedge would have coordinates satisfying both $x + y \leqq 12$ and $x - y \leqq 8$.

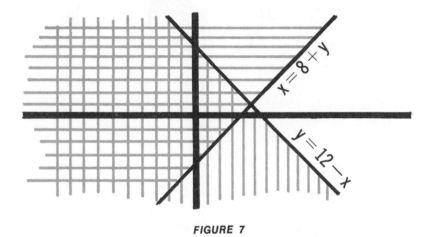

FIGURE 7

If we further require that $x \geqq 0$, then the set of points restricted within or on the boundary of the triangular region in the wedge and to the right of $x = 0$ is the set of points in the intersection of

$$\left\{ \begin{array}{c} x + y \leqq 12 \\ x - y \leqq 8 \\ x \geqq 0 \ . \end{array} \right.$$

In general, it can be seen that the intersection of closed half planes is enclosed within or on the boundary of a convex polygon, and for that reason, is called a "convex polygonal set." In our discussion so far, we have presented only such sets of points in connection with some inequalities drawn out of the thin air. But they are more than a mathematical curiosity; convex sets can lend themselves to practical applications. As the following illustration, quoted with permission of the author, G. Baley Price ("Progress in Mathematics and Its Implications for the Schools"), clearly shows, conditions can arise in a very natural manner restricting the solution of a problem to a set of points within a polygonal region:

> A certain manufacturer has warehouses W_1, W_2 and W_3 which contain 100, 200 and 100 tons, respectively, of his product. The manufacturer receives an order for 125 tons of his product from market M_1 and an order for 225 tons from market M_2. The freight rates from the warehouses W_1, W_2 and W_3 to market M_1 are respectively 1, 2 and 3 dollars per ton; and the freight rates from W_1, W_2 and W_3 to M_2 are respectively 6, 5 and 4 dollars per ton. How many tons should the manufacturer ship from each warehouse to each market to fill the two orders? (*See Figure 8.*)
>
> Let x, y and z denote the number of tons to be shipped from W_1, W_2 and W_3 respectively to M_1; and let u, v and w denote the number of tons to be shipped from W_1, W_2 and W_3 respectively to M_2. Then from the statement of the problem we obtain the following equations and inequalities:

$$x + y + z = 125, \qquad u + v + w = 225,$$
$$x + u \le 100, \qquad y + v \le 200, \qquad z + w \le 100.$$

Finally, if C denotes the total freight charges for making the shipments, then $C = x + 2y + 3z + 6u + 5v + 4w$. The solu-

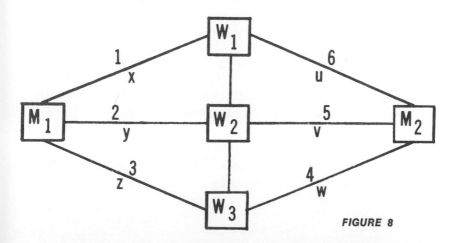

FIGURE 8

tion of the problem is obtained by finding the values x, y, z, u, v and w which satisfy the five equations and inequalities and which give C its minimum value.

The problem of getting either a maximum or a minimum value of a linear function (such as C) when the function is restricted to a convex set is a problem in the study of linear programing. Techniques and theorems have been developed in this study within the last twenty years. Since we do not have these theorems at our disposal here, the following solution will be rather lengthy.

From the statement of the problem, we have the above equation for C which can be expressed as follows on rearranging terms:

$$C = (x+y+z) + (y+2z) + 4(u+v+w) + (2u+v)$$
$$= (x+y+z) + 4(u+v+w) + (y+v) + 2(z+u).$$

Substituting here, the values we have above for $x+y+z$ and for $u+v+w$, we have

$$C = 125 + 900 + (y+v) + 2(z+u).$$

Since $y+v \leq 200$, we have

$$C \leq 125 + 900 + 200 + 2(z+u).$$

As z and u are the number of tons to be shipped to the markets, neither of these can be negative, so the minimum for C is when $z = u = 0$. From this, we have the following:

$$x+y+z = x+y = 125, \qquad u+v+w = v+w = 225,$$
$$x+u = x \leq 100, \qquad z+w = w \leq 100,$$
and $C = 125 + 900 + (y+v) + 2(z+u) = 1025 + (y+v)$.

We are now faced with the problem of the minimum value for C if $C = 1025 + (y+v)$. We may be tempted to suggest that this will be when $y = v = 0$, but we must remember that we are restricted to a convex set, and one of the *constraints* (restrictions) is $x \leq 100$. Another constraint is $x+y = 125$. Together, this gives us a minimum value for y to be 25. Graphically, the minimum value for y is at vertex given by the intersection of the strip from $x = 0$ to $x = 100$ with the set of points determined by $x+y = 125$. Similarly, since

$w \leq 100$ and $v + w = 225$, the minimum value for v is 125. So C has minimum value when $y = 25$ and $v = 125$. Reorganizing all the information we now have,

$$x = 100, y = 25, z = 0, u = 0, v = 125, \text{ and } w = 100$$

will give the minimum value for C.

In the old days, this simple problem would have been solved by trial and error. Nowadays problems are much more complex and may not be as easily solved by guess work. This is where the techniques and theorems of linear programing come to the aid of the mathematician and help him solve the problems of industry.

21

THE ABACUS AND THE SLIPSTICK

ON SEVERAL OCCASIONS, I have seen a standard-size abacus (*see Figure 1*) mounted in a glass atop a $300,000 computer "for use in case of emergency." Now this optional equipment might have been installed as a joke, but it is just as appropriate as grandfather's portrait over the mantel. Although it strictly serves more as a good recorder of calculations than it serves as a true calculator, the abacus (or counting board) is one of the forefathers of the computers in that it signifies an attempt at making computations easier.

FIGURE 1

Most of us will recall that at some point in our lives (around the kindergarten years) we were taught arithmetic from the counting board. The Russians called it a *tschoty,* the Chinese call it a *suan pan,* and the Japanese call it a *soroban.* Whatever it is called, it clearly illustrates that $1 + 3 = 4$ when the shifting of 1 bead followed by the shifting of 3 more results in a total of 4 beads shifted.

People who are dependent upon these tools must develop a

technique for processing the arithmetic operations on these boards. For example, a schoolboy in China, in addition to learning the addition and multiplication combinations, must commit to memory rhymes that he recites (mentally or orally) telling him what to do in any particular situation—when to shift, how to shift. For example, in division by 3, he might repeat to himself, "3 with 2 is changed into 6 and 2." Essentially this tells him that if he is dividing 3 into 2, to change the counters to show a 6 with remainder 2. Behind this mystical prescription, of course, is the concept that the division $2 \div 3$ is treated as $20 \div 3$ (worry about the decimal point later). Obviously, such procedures can be quite cumbersome; nevertheless, similar procedures have been worked out so that even familiar barnyard puzzles (involving some form of equations known as Diophantine equations), requiring solutions to be integers, can be tackled on the abacus. By familiar barnyard puzzles, I mean puzzles of the type:

> Farmer Jones paid $100 for 100 animals. He paid $5 for each horse, $3 for each pig, and 50¢ for each chicken. How many of each kind of animal did he buy if he buys at least one of each?

Aside from the need to memorize endless recipes, this kind of reckoning is limited. But the abacus does represent one attempt at making calculations easier. Other attempts have been made, and vestiges of these are still much in use. Let's try creating some simple machines from scratch.

The creation of our simple adding machine will be an adaptation of the abacus. Instead of beads for counters, we shall use divisions on two sticks A, B. Choose a convenient unit and mark off stick A with this unit, labeling the marks 0 and 1. Now place matching marks on stick B as in Figure 2. The 0 point on either

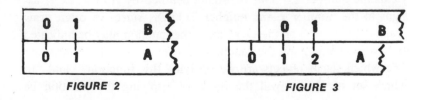

FIGURE 2   **FIGURE 3**

stick will be called the index of the stick. Remember that this is to be an adding machine; so if we slide stick B so that the index on B is matched with the 1 on A, then the 1 mark on B should be pointed at the result of "adding" one unit to one unit. Thus the location on A lined up with the 1 on B should be labeled 2 (*see Figure 3*). Continuing this process, we can add 1 to any whole number that can be represented on A. The final step is to match the markings of A with B, and our adding machine is finished unless we want finer divisions of the unit. This is so simple that one is likely to wonder why all the fuss. Clearly we could have decided right at the start to mark the sticks with the help of a compass or divider with a fixed radius. In the development to follow, the preference for the above procedure should become clearer.

The reader will recognize immediately that the instrument that we have just constructed has all the earmarks of a slide rule; this observation is indeed accurate. This slide rule of ours simply "adds" (or subtracts) chunks of segments just as the abacus "adds" (or subtracts) beads; it just hasn't gotten to the stage yet where it can multiply and divide as its cousin, the conventional slide rule, is supposed to be capable of doing. Technically speaking, of course, neither the conventional slide rule nor the one we just made adds, multiplies, et cetera. The slide rules are merely conveniently marked sticks that interpret tables of combinations such as addition, multiplication, sine, cosine, et cetera, onto the sticks so that corresponding values may be read off. The mathematician calls each of these a *function* in the sense that each number on one of the scales is assigned exactly one number on the other.

As an example of a simple function, we can take one that automatically doubles a given number. Given the number 3, for example, the function assigns the number 6; given the number 7, the function yields the number 14; given the number x, the function yields the number $2x$. This is usually denoted by $f(x) = 2x$ (f assigns to the number x, the number $2x$). Or, since we often draw graphs of functions on the (x,y) axes, the function may be indicated as $y = 2x$. Underlying the function concept are questions such as: "To which set of objects are we applying this function?" and "In which set of objects will the result of applying the function be found?" These are vital questions of *domain* and *range* of a function; they need to be clearly spelled out. For our purpose, we shall assume that both of these sets of objects consist of all numbers, positive, negative, zero, et cetera (in other words, x and y are real numbers) and continue with our task of building machines to do arithmetic after inspecting a few more examples of simple functions.

As a second example of a simple function, we may take one that automatically increases every number by 2. This is the function $f(x) = x + 2$. A more fancy function is one that takes two numbers such as $x, y,$ and produces from them a number called their sum. This is the function $f(x,y) = x + y$. A still fancier function is one that takes the sum $f(x) + f(y)$ and produces $f(xy)$; that is $f(x) + f(y) = f(xy)$.

This last function is what we are seeking for the purpose of building a machine that multiplies. In the process of sliding the sticks, we have seen that we can "add" segments; for a "multiplying" machine on marked sticks, we want markings that give the results of multiplication as the segments are being "added." The object of this quest has already been found by John Napier (1550–1617), through his invention of the *logarithmic* function. Napier had successfully found a way of reducing the task of multiplication to simple addition by the use of this function. Although the function that we shall use as an example below is what is called the common logarithm (to base 10) rather than the Naperian logarithm, the principle is the same. The common logarithm is an adaptation of Napier's logarithm and was produced by Henry Briggs (c. 1560–1630).

First, let's pause to see what kind of task multiplying two numbers such as 2,682,000,000 and 34,470,000 might involve. Each of these numbers in the multiplication may be written in scientific notation thus:

$$2{,}682{,}000{,}000 = 2.682 \times 1{,}000{,}000{,}000 = 2.682 \times 10^9$$
$$34{,}470{,}000 = 3.447 \times 10{,}000{,}000 = 3.447 \times 10^7$$

where each is expressed as a number (factor) multiplied by some power of 10 and in which the factor is a number between 1 and 10 (excluding 10). Then the product

$2,682,000,000 \times 34,470,000 = (2.682 \times 3.447) \times (10^9 \times 10^7)$. The part involving the powers of 10 is easily calculated by the "law of exponents"—namely, $x^m \cdot x^n = x^{m+n}$. Thus, $10^9 \times 10^7 = 10^{16}$. However, the task in the multiplication 2.682×3.447 still has not been made any easier. If each of these can also be expressed as a power of 10, then all that would remain would be to add the exponents. (After that, we need to retranslate this power of 10 in terms of the standard way of writing numbers.) So the job now is turned to finding numbers a, b, such that

$$10^a = 2.682 \quad \text{and} \quad 10^b = 3.447.$$

These are the common logarithms of the numbers 2.682 and 3.447. Notice that since $2682 = 268.2 \times 10 = 26.82 \times 10^2 = \ldots$, one does not need to calculate separately the logarithm for each of these. If $2.682 = 10^a$, then $26.82 = 2.682 \times 10 = 10^a \cdot 10 = 10^{a+1}$, $268.2 = 10^{a+2}$, $\ldots$ Once the a is found for 2.682, the same a with some minor adjustments can be used for 26.82, 268.2, or 2,682,000,000. This shows how handily our use of scientific notation works in with the use of logarithms.

We can now adapt the above ideas to the construction of our multiplying machine. The starting point of our slide rule will be labeled "1" to agree with the smallest factor in the scientific notation. What we shall do now is to decide on how large a segment from 1 to 2 should be. This choice is entirely up to us, but we shall see that once this decision is made, everything else on the rule is determined. So, as in the addition slide rule, we shall mark off two points on A and two matching points on B. These will be labeled 1 and 2 as shown in Figure 4.

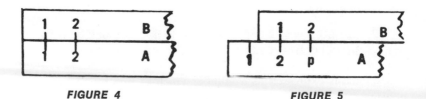

FIGURE 4   **FIGURE 5**

When stick B is shifted so that the index of B is in alignment with the 2 on A, the sticks will be in the position shown in Figure 5. The question then is, "What should we label the point p on A that is now in alignment with the 2 on B?" If this is to be a multiplying machine, then since the B index is at 2 (on A), and p is lined up with 2 (on B), p should be labeled $4 (= 2 \times 2)$. Having done this, the index of B is then shifted over the 4 on A; the point on A that is now in alignment with the 2 on B should then represent the product 2×4. Thus, we continue marking equally spaced points 1, 2, 4, 8, 16, . . . Notice that we have enough intervals by the time we have determined the location for 16, since the point represents the factor in scientific notation and the factor in this notation is a number between 1 and 10.

We have now located points on stick A which we labeled 1, 2, 4, 8, 16. Matching marks similarly labeled on B provide us with a limited tool for multiplying. We should determine some of the in-between numbers if the rule is to be of greater use to us.

To begin, we see that at the end of 1 segment, the label is 2; at the end of 2 segments, the label is 4; . . . To complete the picture, at the end of 0 segments, the label is 1. We have thus the following pairing of numbers:

number of segments:	0	1	2	3	4
marking on scale:	$\mathbf{1} = 2^0$	$\mathbf{2} = 2^1$	$\mathbf{4} = 2^2$	$\mathbf{8} = 2^3$	$\mathbf{16} = 2^4$

This pairing describes a function that we can plot in a graph as in Figure 6, and the function can be expressed as $f(x) = 2^x$. The function can be proved to meet all the requirements of what is called a *continuous* function. This involves the notion of neighborhoods (*see* "A Walk in the Neighborhood"). With an assurance that the function is continuous, we shall connect the plotted points by a continuous curve as is done in the graph. The question as to how many segments from the starting point (index) of A should a marking for a number y be located can now be estimated by finding the x value corresponding to this y. Suppose we want to locate 3 on A or B. Locate 3 on the y axis, go horizontally to find the point on the graph at which $y = 3$, and read the corresponding x value. This is

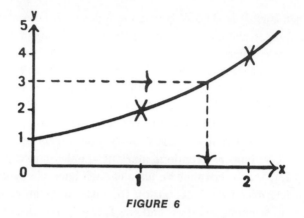

FIGURE 6

the x such that $2^x = 3$. Similarly, other intermediate values can be estimated and marked on the A and B scales by copying the segments on the x axis.

Having seen this, we can then cut strips from graph papers that come already printed with the logarithmic scale and paste these strips onto our sticks for a more finished-looking slide rule. If we had started with the finished product without stopping to examine the details of our primitive slide rule, we might have missed altogether the reason why our machine works. Knowing why our machine works may lead us to designs of other machines calling for other functions. Moreover, with the finished rule on hand now, we can also make a closer examination into how the subdivisions of one segment are related to the subdivisions of another segment on the rule. This is related to the story of biological growth, for example. Also, we might look into the actual computation of the logarithm—an interesting subject in itself.

22

OF MAPS AND MATHEMATICS

THERE ARE MANY BEAUTIFUL maps printed every year, but perhaps the most satisfactory from the mathematical point of view are the quadrangle maps produced by the United States Geological Survey. The Survey has divided the country into a network of quadrangles. The resulting scales, ranging from about one half mile to one mile to the inch, are far larger than those of maps normally found in atlases or maps distributed by gasoline stations. In fact, the scale is so large, that it takes more than two hundred maps to cover a state as small as Maryland.

The quadrangle maps are things of great beauty. Roads, railroads, stations, airports, buildings are printed in black. Bodies of water—such as rivers, ponds, reservoirs, and marshes—are printed in blue. Red is employed for major highways. A light green overlay is used to indicate woodland areas. Brown is used for contour lines.

The amount of information that such a map packs onto one sheet of paper is amazing. If you study the quadrangle map for your own area, you will find the names of hills, streams and other natural features that you little suspected had any name at all. You will find interesting man-made features such as beacons, power lines, canals, abandoned mines and boundary markers. If you live in an area that is not too densely populated, there is a reasonable chance you will find your own house marked as a small black square. On rainy days, these maps can convert you into an armchair explorer. On pleasant days, you can make expeditions to the various surveyors' bench marks, and while on the trip you can try to relate what the eye sees with the symbols indicated on the map. If you are spending your vacation in an unfamiliar part of the country, take along a quadrangle map—you'll be pleased to learn how the intimate knowledge of your whereabouts can increase enormously the fun of staying there.

From the mathematical point of view, the quadrangle maps are satisfying because the scale is large, the geometric relationships between the various features are accurate, and because the maps contain an elaborate system of lines known as *contour lines*. The contour lines establish the altitude of each point on the map. In this way, the sheet of paper, which is a two-dimensional surface, is made to symbolize the bumpy three-dimensional surface of the earth.

Since the map is an accurate but scaled-down version of what is to be found on the land, the angles between objects on the map must be the same as between their real counterparts. Let me show how this fact can be employed, by relating an experience of my own. I recently spent several weeks at Echo Pond, Vermont. This pond is near the Canadian boundary line and nestles among a number of small mountains. The region was new to me, but I had written in advance for quandrangle maps, as I wanted to identify the natural features of the place. The pond is about a mile across at its widest, and when I stood on the shore, I saw the mountains at various distances and heights overlapping one another, presenting a series of corrugations to the eye.

The cabin was at the water and was actually indicated on the quadrangle map. I located it from its relative position along the pond road, and from the shape of the private drive leading from the road to the cabin. A small stream fed into the pond ten feet from our landing. This was also on the quadrangle map, and I felt that this really pinpointed my location on the shore.

I spread the map on a table at the shore, and stuck it to the table by a thumbtack through my location. The map was now free to rotate about this position. To be of any use, I would have to fix the direction of the map. I had no compass and would therefore have to get a fix by associating one feature on the landscape with its counterpart on the quadrangle map.

There was one peak on the horizon noticeably higher than all the others. I examined the map to a distance of about seven miles in the direction of the opposite shore, and found that it showed Bald Mountain, with an altitude of 3,315 feet, as the tallest mountain in the vicinity. The map indicated a fire tower on the top of this mountain. If I looked hard enough, I could just barely see a tiny object on the top of the peak I had tentatively identified as Bald Mountain. I therefore felt sure of the identification. I turned the map about the tack until the line drawn from the tack to the Bald Mountain fire tower on the map coincided with the imaginary line drawn between me and the real fire tower. The map was now in proper position, and I secured it with further tacks.

The whole landscape now fell into position. Each line of sight was properly indicated on the map. In no time I had identified Elan Hill, Bear Hill, Tripp Hill, Job's Mountain, and was able to read off their respective distances and altitudes. The clump of houses at the stream connecting Echo Pond and Seymour Lake was at the proper angle. The abandoned farmhouse across the pond was in the right spot. It stood in a clearing, and the cleared and wooded areas were accurately shown on the quadrangle map.

You can use the quadrangle maps to identify the features in any area of interest. All that is required is to locate yourself on the map, and then locate one additional object on the landscape to fix the direction. In order to be accurate, the distance between yourself and the object should be several miles long. A wide vista is therefore required.

The study of maps can lead to some of the most intricate theory in the whole of mathematics. But there are many simple, yet interesting, mathematical projects that can be carried out with quadrangle maps.

Finding the Area of a Pond or Lake

Trace the outline of the pond onto a piece of graph paper. Count the number of squares that lie inside the contour, estimating the contribution of the partial squares at the boundary. Multiply

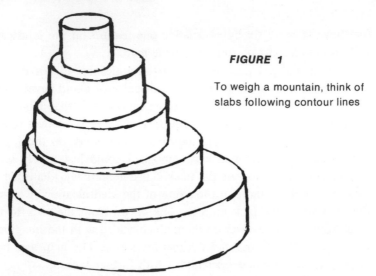

FIGURE 1

To weigh a mountain, think of
slabs following contour lines

this total by the proper scale factor to obtain the area of the pond.
More sophisticated procedures can be developed; in fact, this is a
fine place to begin a study of the topic that is known as *approximate integration.*

Finding the Length of a Stream

A crude procedure is to place a sequence of close points on
the stream and to regard the stream as a broken line with these
points as vertices. Measure the lengths of the individual segments
and add them up. For a more accurate procedure that is useful
in working with maps, read pages 108–13 of Hugo Steinhaus'
Mathematical Snapshots.

Weighing a Mountain

Regard the mountain as composed of slabs (*Figure 1*). The
bottom of each slab is a contour line of the mountain, and the sides
of the slab are perpendicular to the bottom. Compute the volume
of each slab by multiplying the area of the bottom by the height.
Estimate the volume of the mountain by summing the volumes of
the slabs. Multiply this total by the density of slate, granite, or whatever material the mountain is composed of. For example, the
density of granite is about 2.7 grams per cubic centimeter.

Studying the Drainage Net of a River

The branching tributaries of a river are called a *drainage net*. River scientists have defined the *order* of a stream in a drainage net. The smallest unbranched tributary in the headwaters is a stream of order 1. A stream whose tributaries are of order 1 is said to be of order 2, and so forth for the higher orders (*Figure 2*). What do you suppose is the order of the Mississippi River? Quadrangle maps can help you find out. For an interesting article that describes drainage nets, order of streams, and some mathematical laws relating to them, read "Rivers," by L. B. Leopold, in the *American Scientist,* December 1962.

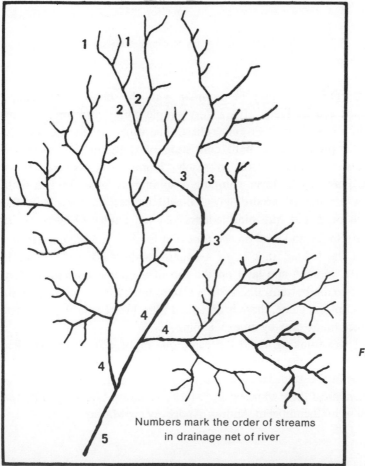

FIGURE 2

Numbers mark the order of streams
in drainage net of river

23

"MR. MILTON, MR. BRADLEY— MEET ANDREY ANDREYEVICH MARKOV"

SOME OF YOU may be familiar with the board game called "Chutes and Ladders." It's a very simple game, and many young children have it in their game chest. The object of the game is to move a piece over a path consisting of a hundred consecutive squares. The first person to reach "home"—the square 100— wins. Players take turns spinning a spinner (*Figure 2*) and their pieces are moved accordingly. To add interest to the game, the path is peppered with nine ladders and ten chutes (*Figure 1*). If you are lucky and land on the foot of a ladder, you advance to the top of the ladder, that is, to a square farther along the path. If you are unlucky and land on the top of a chute you immediately plunge to the bottom, that is, to a square back along the path. Near the end of the game, there is an added fillip in the rule that you can advance home only by an exact spin.

I was thinking about the game one night, wondering whether either you or the president of the Milton-Bradley Company, the makers of the game, realize that the game is a fine example of the mathematical theory known as "Markov Chains," developed by the Russian mathematician Andrey Andreyevich Markov.

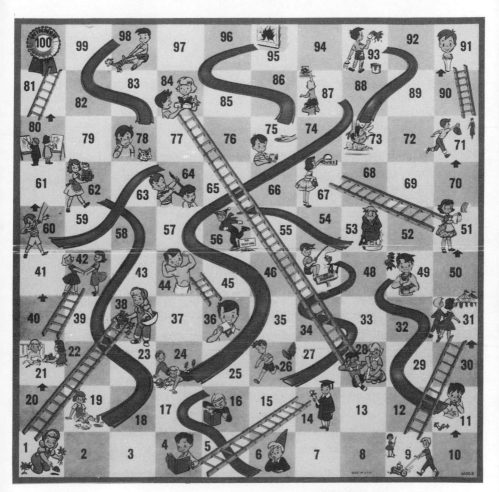

FIGURE 1

FIGURE 2

After I fell asleep, I dreamed a dream that took place in the research and development laboratories of the Milton-Bradley Co. in Springfield, Massachusetts.

Mr. Milton and Mr. Bradley are talking together, sitting over a Chutes and Ladders board.

MR. MILTON. This new game of ours seems like good fun. But we ought to test it out before putting it on the market.

MR. BRADLEY. That's right. It's for small children, which means each game should not last too long, on the average. A child's attention span is not too great.

MR. MILTON. But the game shouldn't be too short either.

MR. BRADLEY. Do you think we've put in enough chutes and ladders? How often do you suppose you get on a chute or on a ladder? Too many chutes might be frustrating. We'd better test it out.

MR. MILTON. But I don't know how to test it. I've asked our bookkeepers, Mr. Diamond and Mr. Spade, to play a few hundred games, and keep records. But they don't want to. They say they only know how to play gin rummy.

MR. BRADLEY. What'll we do? What'll we do?

Enter ghost of Andrey Andreyevich Markov. Ghost approaches table where Milton and Bradley are sitting dejectedly in front of the game. He speaks to the two men, using the dialect of Russian ghosts who speak only printed English.

A. A. MARKOV. What is great lamentation here? What is such terrible problem?

MR. MILTON. Huh?

A. A. MARKOV. Aha! Very, very interesting game. Is example of my beautiful chains. Tsyepee Markova.

MR. BRADLEY. Who are you? Why are you intruding in our private office?

A. A. MARKOV. My pleasure to introducing you to myself, Andrey Andreyevich Markov, mathematician and probability expert. This game Choots and Letters you are invented is fine example of my chains!

MR. BRADLEY. Chains? What chains? I don't see any chains.

A. A. MARKOV. You are making inquirings about average length of game? With my chains, is very easy to compute such inquirings. And many more similar such inquirings.

MR. MILTON. Maybe he can help us out, Bradley.

A. A. MARKOV. Absolutely certain. Is very simple. Take truncated transition matrix and subtract him from identity. Find inverse to difference, and making sum of elements of first row of inverse. Is clear?

MR. BRADLEY. Chains? Matrix? Inverse? Elements?

MR. MILTON. Transition? Truncated? Identity? What?

A. A. MARKOV. Do sveedanya. Seeing you keeds again. Remember secret of my chains.

The dream ends as the ghost exits, rubbing his hands in satisfaction, while Messrs. Milton and Bradley exit in considerable puzzlement.

What is the explanation of these mysterious goings on? Andrey Andreyevich Markov (1856–1922) was a world-famous Russian mathematician and professor at the University of St. Petersburg. He made many discoveries in the theory of numbers and in differential equations. But it is his work in probability that concerns us here. In 1907, Markov studied systems of objects that change from one state to another state, according to certain laws of probability. When an initial state can lead to subsequent states, and these states, in turn, can lead to further states, then we have what Markov called a *probability chain,* or what the world of science now calls a Markov chain, in his honor.

The game of Chutes and Ladders will help us understand these ideas. Let us play a solitaire version of the game. At the beginning, our piece is off the board. We will call this condition "state 0." We now spin the spinner for the first time. According to the way the spinner is constructed, the numbers 1, 2, 3, 4, 5 and 6 will be spun with equal likelihood. In the language of probability, the probability of spinning either 1, 2, 3, 4, 5 or 6 is 1/6. This means that, with probability 1/6, we move our piece to one of the squares numbered 1, 2, 3, 4, 5 or 6 after the first spin.

FIRST TWO SPINS OF SOLITAIRE GAME OF "CHUTES AND LADDERS"

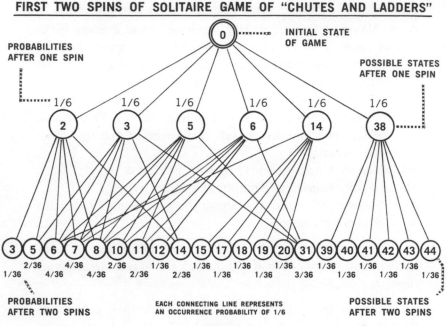

FIGURE 3

Diagram shows all possibilities after two spins. Mathematician A. A. Markov described the probability links as a "chain."

If we are lucky and spin a 1, we go to 1, but then we move up to the ladder to 38 without an additional spin. If we are lucky and spin a 4, we go to 4 and move up a ladder to 14.

The result of one spin can be summarized as follows. We move from state 0 to state 2, 3, 5, 6, 14 or 38 with equal probability of 1/6. (This is the top part of the pyramid in Figure 3.) We will not consider squares 1 and 4 as states because our piece does not come to rest there. By the same token, we will not consider as states the bottom of any ladder or the top of any chute.

Next, consider the second spin. As we make the spin, the game can be in one of six states, and we must consider the possibilities separately. Suppose, for example, that the first spin leaves us in state 2. On the spin, 1, 2, 3, 4, 5, 6 show up on the dial with equal probability. Hence we may move to square number 3, 4, 5,

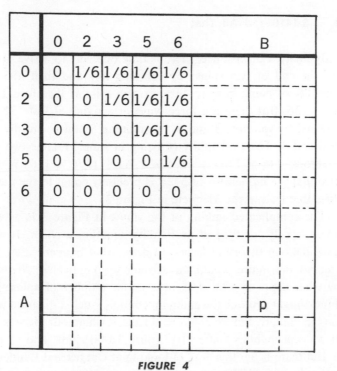

	0	2	3	5	6		B	
0	0	1/6	1/6	1/6	1/6			
2	0	0	1/6	1/6	1/6			
3	0	0	0	1/6	1/6			
5	0	0	0	0	1/6			
6	0	0	0	0	0			
A							p	

FIGURE 4

Table lists all probabilities (*p*) of passing from one state to another in game of "Chutes and Ladders." Table specifies entire game, is called Matrix of Transition Probabilities.

6, 7 or 8. But 4 is the bottom of a ladder. If we land there, we immediately ascend to 14. Hence, starting from state 2, we may move to states 3, 5, 6, 7, 8 or 14 with equal probability. (This is illustrated by the left-hand portion of the bottom layer of the pyramid in Figure 3.)

To describe all the possibilities on the second spin, we must consider all the possibilities on the first spin. To go through one more possibility, if we were in state 14 at the end of the first spin, we could move with equal probability to squares 15, 16, 17, 18, 19 or 20. But if we were unfortunate and landed on 16, we would plunge to 6. Hence, from state 14 we move, with equal probability, to state 6, 15, 17, 18, 19 or 20.

All the possibilities after two spins are described in Figure 3. The game can be in a total of 21 states. But these states are not equally likely. For instance, there is only one way to arrive at state 44 at the end of two spins: by spinning a 1 and then a 6. The probability of being in state 44 after two spins is therefore $1/6 \times 1/6 = 1/36$. But there are three ways of arriving at state 31 after two spins: by spinning 3 and 6, by spinning 5 and 4, and by spinning 6 and 3. The probability of being in state 31 after two spins is therefore $(1/6 \times 1/6) + (1/6 \times 1/6) + (1/6 \times 1/6) = 3/36$. Notice that the numerator of $3/36$ is precisely equal to the number of lines that lead to the 31 circle in Figure 3.

The complicated linking of the states in Figure 3 is what led Markov to call this type of probability situation a *chain*. It would be possible—in theory at least—to draw what happens after three spins, but the figure would be extremely complicated. We could draw what happens after 4, 5 or a larger number of spins. Since bad luck might prolong the game indefinitely, such a diagram would go on indefinitely, but we could read from it where our piece is (and with what probability) after 100 spins, 1,000 spins, and so forth.

But there is another way of looking at Chutes and Ladders—a way that is more algebraic and less geometric. The game has a total of 82 states (100 squares − 9 ladder bottoms − 10 chute tops + the initial zero state). If the game is in a certain state, then it proceeds in one spin to a given state with a certain given probability. If the transition is impossible (such as moving from state 0 to state 8), the probability is 0. We can therefore make a table that lists all the probabilities of passing from one state to another state, and the whole game is completely specified by this table. In the case of Chutes and Ladders, the table would have $82 \times 82 = 6,724$ entries in it, most of them zero. This table is known as the *matrix of transition probabilities*—"matrix" being the common mathematical term for a rectangular array of numbers, and "transition" referring to the transition of the game from one state to another. Figure 4 shows a portion of the transition matrix for Chutes and Ladders.

In the theory of Markov chains, the transition matrix plays a fundamental role. Mathematicians in the years following Markov have discovered formulas, expressed in terms of this matrix, for many statistical quantities associated with probability chains. One such formula was given in the playlet at the beginning of the article.

The algebra of matrices has been most important in promoting these discoveries.

Markov chains have found application in a wide variety of sciences, including biology, physics, astrophysics, chemistry, operations research, and psychology. It would have been hard for Markov to foresee that electronic computers would one day work out probabilities in these sciences in the manner indicated by him. It would have been equally hard for him to foresee that many probability situations are so complex that they defy his analysis, and that these problems are solved in the manner offhandedly suggested by Mr. Milton: by playing games. But the games are played silently and rapidly in the circuits of electronic computers.

24

3.1416 AND ALL THAT

I BELONG TO THE ANCIENT and Honorable Society of Pi Watchers. The function of this society is to promote, to collect and to disseminate information about the number pi. There are no formal meetings, no newsletters or regular publication of minutes. On the other hand, anyone can join. New members of the society might very well read my status report on pi, which was written in 1961 and can be found in Chapter 17 of *The Lore of Large Numbers*. Another recent document of prime importance to pi watchers is the article by D. Shanks and J. W. Wrench, Jr., entitled "Calculation of π to 100,000 Decimals," which can be found in the January 1962 issue of *Mathematics of Computation*.

For the reader who needs his memory refreshed just a bit, the number pi (generally written as the Greek letter π) is the ratio of the circumference to the diameter of a circle. In decimals we have

$$\pi = 3.14159 \quad 26535 \quad 89793 \quad \ldots$$

so that the number in the title of this article—the popular value of π—is slightly in error. The value of π cannot be expressed exactly as the ratio of two integers; this means that π is an irrational number. A consequence of this is that when π is expressed in decimal form, its decimal digits can never repeat in a pattern. But more than this is true; π is a *transcendental* number. This means that π cannot satisfy an algebraic equation with integer coefficients. Thus, for example, it is impossible for π to satisfy the equation

$$\pi^4 - 240\pi^2 + 1492 = 0$$

or, in general, an equation of the form

$$a_0\pi^n + a_1\pi^{n-1} + a_2\pi^{n-2} + \cdots + a_{n-1}\pi + a_n = 0$$

where $a_0, a_1, \ldots, a_n$ are integers and n is a positive integer. This impossibility puts to rest all the attempts at circle squaring by means of certain elementary operations with the ruler and compass.

Though the decimal digits of π are perfectly determined, and with sufficient patience we may presumably compute as many of them as we like, it has been conjectured that these digits are completely random. This is to be understood in the sense that each of the digits $0, 1, 2, \ldots, 9$ occurs on the average $1/10$ of the time; each of the combinations $00, 01, \ldots, 99$ occurs on the average $1/100$ of the time; each of the triples $000, 001, 002, \ldots, 999$ occurs on the average $1/1000$ of the time; et cetera. This conjecture has not been proved as yet. It appears to be remarkably difficult to establish, and it constitutes the principal unsolved problem about the nature of the number π.

One of the facts that strikes π watchers immediately is that there has been a drive to compute π to more and more decimal places. This drive has been going on for thousands of years. The direct utility to problems of measurement or engineering of ultra-high-accuracy computations of π is nil. What, therefore, is behind this drive? Many things; and not the least of them is the fact that there are many people in this world who like to break records, and many who love to compute just for the sake of computing. Whereas some people like to spend their leisure time playing bridge or reading mystery stories, there are others who are never so happy as when they are working with pencil, paper and a computing machine. (Remote computer input via home telephone lines should soon increase their joy.)

But there have been more theoretical reasons for computing high-precision values of π. Prior to 1761, when Lambert just established that π is an irrational number, a reason for computing π to many figures was the hope that such a calculation might show periodic patterns of its decimal digits and hence its fractional form would stand revealed. In recent years, many people, including one of the foremost of American mathematicians, the late John von Neumann, have interested themselves in the distribution of the decimal digits of π. Statistical analysis of the many digits of π produced by computers tends to confirm the conjecture that the digits are random.

The electronic digital computer has thus far multiplied our ability to compute the digits of π by a factor of a thousand. Here are some of the milestones in the computer race toward a million decimals.

>   1949, Aberdeen Proving Ground, U.S.A., 2,036 decimals
>   1954, Watson Scientific Laboratory, U.S.A., 3,093
>   decimals
>   1959, IBM 704, F. Genuys, France, 10,000 decimals
>   1959, Ferranti Pegasus, Felton, England, 10,007 decimals
>   1961, Shanks and Wrench, U.S.A., 100,000 decimals

The most recent records in the π race are French. In a report dated February 26, 1966, put out by the Commissariat à l'Énergie Atomique in France, Jean Gilloud and J. Fillatoire give 250,000 decimals of π. The calculation was performed on an IBM 7030 (STRETCH) machine.

Just one year later, on February 26, 1967, a second report "500,000 Décimales de π" was written by Jean Gilloud and Michèle Dichampt. The calculation was performed on a CDC 6600. The total running time, including conversion of the final result to decimal notation and the check, was 44 hours, 45 minutes. Figuring $200 as a reasonable (American) price for computing time, this computation cost almost two cents per decimal place, exclusive of programing costs.

The computation and check employed the trigonometric formulas

$$\pi = 24 \tan^{-1} \frac{1}{8} + 8 \tan^{-1} \frac{1}{57} + 4 \tan^{-1} \frac{1}{239}$$

$$\pi = 48 \tan^{-1} \frac{1}{18} + 32 \tan^{-1} \frac{1}{57} - 20 \tan^{-1} \frac{1}{239}$$

The inverse tangent functions are given by the infinite series

$$\tan^{-1}x = x - \frac{x^3}{3} + \frac{x^5}{5} - \frac{x^7}{7} + \frac{x^9}{9} - \cdots$$

As of the date of writing (August 15, 1968) this is the best result known to the writer. However, a million decimals are expected momentarily.

Thus far the principal contenders in the π race have been Americans, English and French. When will the Russians enter this race? When the Russian Mathematical Bear rouses himself and plunges into the π pool, there will be a real splash, for the Russians are notorious for doing things on the grand scale. Will it be one million digits? Five million? Ten million?

To produce π on an electronic computer to superaccuracy is not an open-and-shut matter. Normally—perhaps 95 per cent of the time—computations are done with about seven or eight decimals. This is called *single-precision* computation. Since roundoff (the dropping of the least significant decimals when arithmetic operations are performed) vitiates the accuracy of certain types of computations, one may be forced to send the computer into a double- or even a triple-precision mode of operation. To compute π to thousands of places requires ultra-poly-precision, and this means efficient formulas to begin with, special clever types of computer programing as well as various tricks to save time and memory space. Over the years, π has been a fertile source of some of the most difficult problems in the whole of mathematics; the electronic computer has added a completely new parcel of problems to its story.

About five years ago I spoke over the phone to one of the foremost π racers in this country. "What is an upper bound for the number of decimal places of π that will *ever* be computed?" I asked him. He thought a moment and then replied, "The human race will never see one billion decimals of π." I have yet to find out whether this figure was based upon his estimate of the future potentialities of computing machines or whether it reflects my friend's feelings about the future of the human race.

It is worth pointing out that 2,000 readable digits fit comfortably on a printed page so that one billion digits represent a thousand large books of five hundred pages each. This is about 200 running feet of stacked books. So much for the dramatics, now what about the mathematics? The above prediction leads us squarely to two questions: (1) Is it possible to develop a method for computing the nth digit of π without having to compute all the digits that precede it? (2) What conceivable meaning can be attached to an infinite decimal when we are not in a position *ever* to tell what its digits are?

So long until the Ancient Society meets once again.

SOME MILESTONES IN THE π RACE

Ptolemy, 150 A.D.	4 decimals
Viete, 1579	10
Romanus, 1593	16
Van Ceulen, 1610	33
Snell, 1621	35
Sharp, 1699	72
Machin, 1706	101
Vega, 1794	137
Dase, 1844	201
Rutherford, 1853	441
Shanks, 1873	707 *
Aberdeen Proving Grounds, 1949	2,036
Watson Scientific Laboratory, 1954	3,093
Genuys, Felton, 1959	10,000
Shanks and Wrench, 1961	100,000
Gilloud and Fillatoire, 1966	250,000
Gilloud and Dichampt, 1967	500,000
???	1,000,000

. . .

---

* 527 correct.

BIBLIOGRAPHY

1. The Problem That Saved a Man's Life

Bell, Eric Temple, *Men of Mathematics*. New York: Simon and Schuster, 1937. Chapter 4 (for Fermat), Chapters 25, 27 (for Kummer).

Dickson, Leonard E., *History of the Theory of Numbers*. New York: Chelsea Publishing Company, 1952. Vol. II, Chapter 26.

Russell, Bertrand, *A History of Western Philosophy*. New York: Simon and Schuster, 1945. Chapter 3 is a delightful essay on Pythagoras' ideas.

Vandiver, H. S., "Fermat's Last Theorem—Its History and the Nature of the Known Results Concerning It," *American Mathematical Monthly*, Vol. 53 (1946), pp. 555–76.

2. The Code of the Primes

Courant, Richard, and Robbins, Herbert, *What Is Mathematics?* New York: Oxford University Press, 1941.

Dantzig, Tobias, *Number—The Language of Science*. New York: Macmillan, 1933.

Freund, John E., *A Modern Introduction to Mathematics*. Englewood Cliffs, N.J.: Prentice-Hall, 1956.

Hogben, Lancelot, *Mathematics for the Million*. New York: W. W. Norton, 1937.

Kline, Morris, *Mathematics—A Cultural Approach*. Reading, Mass.: Addison-Wesley Publishing, 1962.

3. Pompeiu's Magic Seven

Birkhoff, Garrett, and MacLane, Saunders, *A Survey of Modern Algebra,* revised edition. New York: Macmillan, 1953. Pp. 26–29.

Davis, Philip J., *The Lore of Large Numbers*. New York: Random House, 1961. Chapter 20.

Hardy, Godfrey H., and Wright, Edward M., *The Theory of Numbers*, 4th Edition. Oxford: Clarendon Press. P. 51.

Pompeiu, Dimitrie, *Oeuvre mathématique*. Bucharest: Académie de la République populaire romaine, 1959.

4. What Is an Abstraction?

Berge, Claude, *The Theory of Graphs*. London: Methuen, 1962.

Ore, Oystein, *Graphs and Their Uses*. New York: Random House, New Mathematical Library, 1963.

————, *Theory of Graphs*. American Mathematical Society Colloquium Publications, Vol. 38. Providence, R.I., 1962.

5. Postulates—The Bylaws of Mathematics

Eves, Howard, and Newsom, Carroll V., *An Introduction to the Foundations and Fundamental Concepts of Mathematics*. New York: Rinehart, 1958.

Mathematical Association of America, Committee on the Undergraduate Program, *Elementary Mathematics of Sets with Applications*. Ann Arbor, Mich.: Cushing-Malloy, 1955.

Meserve, Bruce E., *Fundamental Concepts of Geometry*. Cambridge, Mass.: Addison-Wesley Publishing, 1955.

Newman, James R., *The World of Mathematics*, 4 vols. New York: Simon and Schuster, 1956.

Schaaf, William L., *Basic Concepts of Elementary Mathematics*. New York: John Wiley and Sons, 1960.

Willerding, Margaret F., and Hayward, Ruth A., *Mathematics—The Alphabet of Science*. New York: John Wiley and Sons, 1968.

6. The Logical Lie Detector

Andree, Richard V., *Selections from Modern Abstract Algebra*. New York: Henry Holt, 1958.

Christian, Robert R., *Introduction to Logic and Sets*. Boston: Ginn, 1958.

Courant, Richard, and Robbins, Herbert, *What Is Mathematics?* New York: Oxford University Press, 1941.

Mathematical Association of America, Committee on the Undergraduate Program, *Elementary Mathematics of Sets with Applications.* Ann Arbor, Mich.: Cushing-Malloy, 1955.

Rosenbloom, Paul C., *The Elements of Mathematical Logic.* New York: Dover Publications, 1951.

7. Number

Aaboe, Asger, *Episodes from the Early History of Mathematics.* New York: Random House, New Mathematical Library, 1964.

Barker, Stephen F., *Philosophy of Mathematics.* Englewood Cliffs, N.J.: Prentice-Hall, 1964.

Bell, Eric Temple, *The Development of Mathematics.* New York: McGraw-Hill Book, 1945.

Dantzig, Tobias, *Number—The Language of Science.* New York: Macmillan, 1954.

Davis, Philip J., *The Lore of Large Numbers.* New York: Random House, 1961.

Isaacs, G. L., *Real Numbers.* New York: McGraw-Hill Book, 1968.

Kline, Morris, *Mathematics and the Physical World.* New York: Thomas Y. Crowell, 1959.

Smith, David Eugene, *History of Mathematics,* Vol. II, *Special Topics of Elementary Mathematics.* New York: Dover Publications, 1958.

8. The Philadelphia Story

Davis, Philip J., "The Criterion Makers: Mathematics and Social Policy," *American Scientist,* Vol. 50 (1962), No. 3, pp. 258A–274A.

9. Poinsot's Points and Lines

Coxeter, H. S. M., *Regular Polytopes.* London: Methuen, 1948. Pp. 93–94.

Davis, Philip J., and Rabinowitz, Philip, *Numerical Integration.* Waltham, Mass.: Blaisdell Publishing, 1967. P. 151. (For Weyl's Theorem and its application.)

Poinsot, Louis, and others, *Abhandlungen über die regelmässigen Sternkörper,* translated and edited by Robert Haussner. Leipzig: W. Engelmann, 1906. (For German version Poinsot's original article of 1810; no English translation is available.)

10. Chaos and Polygons

Aaboe, Asger, *Episodes from the Early History of Mathematics*. New York: Random House, 1964.

Cundy, H. Martyn, and Rollett, A. P., *Mathematical Models*. London: Oxford University Press, 1951.

Klein, Felix, *Famous Problems of Elementary Geometry*. New York: Chelsea Publishing, 1955.

Kline, Morris, *Mathematics—A Cultural Approach*. Reading, Mass.: Addison-Wesley Publishing, 1962.

Olds, Charles D., *Continued Fractions*. New York: Random House, 1962.

11. Numbers, Point and Counterpoint

Andree, Richard V., *Selections from Modern Abstract Mathematics*. New York: Henry Holt, 1958.

Courant, Richard, and Robbins, Herbert, *What Is Mathematics?* New York: Oxford University Press, 1941.

Kline, Morris, *Mathematics—A Cultural Approach*. Reading, Mass.: Addison-Wesley Publishing, 1962.

Van der Waerden, B. L., *Science Awakening*. New York: John Wiley and Sons, 1963.

12. The Mathematical Beauty Contest

Hilbert, David, and Cohn-Vossen, Stefan, *Geometry and the Imagination*. New York: Chelsea Publishing, 1952. Chapter 4.

Kazarinoff, Nicholas, *Geometric Inequalities*. New York: Random House, New Mathematical Library, Vol. 4, 1961. Chapter 2.

Steinhaus, Hugo, *Mathematical Snapshots*. New York: Oxford University Press, 1960. Pp. 159–64.

13. The House That Geometry Built

Bing, R. H., *Elementary Point Set Topology*, Herbert Ellsworth Slaught Memorial Papers No. 8. Menasha, Wisc.: Mathematical Association of America, 1960.

Birkhoff, Garrett, and MacLane, Saunders, *A Survey of Modern Algebra*, revised edition. New York: Macmillan, 1953.

Courant, Richard, and Robbins, Herbert, *What Is Mathematics?* New York: Oxford University Press, 1941.

Eves, Howard, and Newsom, Carroll V., *An Introduction to the Foundations and Fundamental Concepts of Mathematics*. New York: Rinehart, 1958.

Meserve, Bruce E., *Fundamental Concepts of Geometry*. Cambridge, Mass.: Addison-Wesley Publishing, 1955.

14. Explorers of the Nth Dimension

Coxeter, H. S. M., *Introduction to Geometry*. New York: John Wiley and Sons, 1961. Chapter 22.

Synge, J. L., "The Geometry of Many Dimensions," *Mathematical Gazette,* Vol. 33 (1949), pp. 249–63.

Ulam, Stanislaw M., *A Collection of Mathematical Problems*. New York: Interscience Publishers, 1960. Chapter 8, "Computing Machines as a Heuristic Aid."

15. The Band-Aid Principle

Kline, Morris, *Mathematics and the Physical World*. New York: Thomas Y. Crowell, 1959. Chapter 25.

Kreyszig, Erwin, *Differential Geometry*. Toronto: University of Toronto Press, 1959.

Lyusternik, Lazar A., *Shortest Paths*. New York: Macmillan, 1963.

Stäckel, Paul G., "Bemerkungen zur Geschichte der geodätische Linien," Sächsische Gesellschaft der Wissenschaften, *Berichte,* Math-Physische Klasse, Vol. 45 (1893), pp. 444–67.

16. The Spider and the Fly

Courant, Richard, and Robbins, Herbert, *What Is Mathematics?* New York: Oxford University Press, 1941.

Eves, Howard, and Newsom, Carroll V., *An Introduction to the.Foundations and Fundamental Concepts of Mathematics*. New York: Rinehart, 1958.

Freund, John E., *A Modern Introduction to Mathematics*. Englewood Cliffs, N.J.: Prentice-Hall, 1956.

Kline, Morris, *Mathematics—A Cultural Approach*. Reading, Mass.: Addison-Wesley Publishing, 1962.

17. A Walk in the Neighborhood

Chinn, William G., and Steenrod, Norman E., *First Concepts of Topology*. New York: Random House, 1966.

Court, Nathan A., *Mathematics in Fun and Earnest*. New York: New American Library of World Literature, 1961.

Dantzig, Tobias, *Number—The Language of Science*. Garden City, N.Y.: Doubleday, 1956.

Kasner, Edward, and Newman, James R., *Mathematics and the Imagination*. New York: Simon and Schuster, 1940.

Meserve, Bruce E., *Fundamental Concepts of Geometry*. Cambridge, Mass.: Addison-Wesley Publishing, 1955.

Patterson, Edward M., *Topology*. Edinburgh: Oliver and Boyd, 1956.

Young, Frederick H., *Limits and Limit Concepts*. Boston: Ginn, 1964.

Zippin, Leo, *Uses of Infinity*. New York: Random House, New Mathematical Library, Vol. 7, 1962.

18. Division in the Cellar

Isaacson, Eugene, and Keller, Herbert B., *Analysis of Numerical Methods*. New York: John Wiley and Sons, 1966.

Kuo, Shan Sun, *Numerical Methods and Computers*. Reading, Mass.: Addison-Wesley Publishing, 1965.

Vilenkin, N. Y., *Successive Approximation*. New York: Macmillan, 1964.

19. The Art of Squeezing

Chinn, William G., and Steenrod, Norman E., *First Concepts of Topology*. New York: Random House, 1966.

Courant, Richard, and Robbins, Herbert, *What Is Mathematics?* New York: Oxford University Press, 1941.

Dubisch, Roy, *The Nature of Number*. New York: Ronald Press, 1952.

Goodstein, Reuben L., *Fundamental Concepts of Mathematics*. New York: Pergamon Press, 1962.

Newsom, Carroll V., *Mathematical Discourses: The Heart of Mathematical Science*. Englewood Cliffs, N.J.: Prentice-Hall, 1964.

Niven, Ivan M., *Numbers: Rational and Irrational*. New York: Random House, New Mathematical Library, 1961.

Ribenboim, Paulo, *Functions, Limits and Continuity*. New York: John Wiley and Sons, 1964.

20. The Business of Inequalities

Hadley, George, *Linear Algebra*. Reading, Mass.: Addison-Wesley Publishing, 1961.

Johnston, John B., Price, G. Baley, and Van Vleck, Fred S., *Linear Equations and Matrices*. Reading, Mass.: Addison-Wesley Publishing, 1966.

Kemeny, John G., Snell, J. Laurie, and Thompson, Gerald L., *Introduction to Finite Mathematics*. Englewood Cliffs, N.J.: Prentice-Hall, 1955.

MacDonald, John Dennis, *Strategy in Poker, Business and War*. New York: W. W. Norton, 1950.

Metzger, Robert W., *Elementary Mathematical Programming*. New York: John Wiley and Sons, 1958.

Price, G. Baley, "Progress in Mathematics and Its Implications for the Schools," *The Revolution in School Mathematics*. Washington: National Council of Teachers of Mathematics, 1961.

21. The Abacus and the Slipstick

Davis, Philip J., *The Lore of Large Numbers*. New York: Random House, New Mathematical Library, 1961.

Niven, Ivan M., *Numbers: Rational and Irrational*. New York: Random House, New Mathematical Library, 1961.

Sawyer, W. W., *Mathematician's Delight*. Baltimore: Penguin Books, 1943.

22. Of Maps and Mathematics

Deetz, Charles H., and Adams, Oscar S., *Elements of Map Projection*. Washington: U. S. Department of Commerce, Coast and Geodetic Survey, 1945. (Obtainable at U. S. Government Printing Office.)

Leopold, Luna B., "Rivers," *American Scientist*, Vol. 5 (December 1962), pp. 511–37.

23. "Mr. Milton, Mr. Bradley—Meet Andrey Andreyevich Markov"

Bazley, N. W., and Davis, Philip J., "Accuracy of Monte Carlo Methods in Computing Finite Markov Chains," *Journal of Research of the National Bureau of Standards,* Vol. 64B, No. 4 (October–December, 1960), pp. 211–15.

Kemeny, John G., and Snell, J. Laurie, *Finite Markov Chains.* Princeton, N.J.: D. Van Nostrand, 1960.

24. 3.1416 and All That

Davis, Philip J., *The Lore of Large Numbers.* New York: Random House, New Mathematical Library, 1961. Chapter 17.

Hobson, Ernest W., *Squaring the Circle.* New York: Chelsea Publishing, 1958.

Shanks, D., and Wrench, J. W., Jr., "Calculation of π to 100,000 Decimals," *Mathematics of Computation,* Vol. 16 (1962).

ABOUT THE AUTHORS

PHILIP J. DAVIS has been a professor of applied mathematics at Brown University since 1963. Before that time, he was a mathematician with the National Bureau of Standards in Washington, D. C. He received both his B.S. and Ph.D. from Harvard University. He is the author of one textbook, two technical mathematics books, and the highly popular, *The Lore of Large Numbers.* He lives in Providence, Rhode Island, with his wife, Hadassah, and their four children.

WILLIAM G. CHINN is also involved in mathematics education. Now teaching mathematics at City College of San Francisco, he was formerly mathematics supervisor of the San Francisco Unified School District. He has authored seven texts for the School Mathematics Study Group and three other books including *First Concepts of Topology.* A graduate of the University of California at Berkeley, he has also received a certificate of meteorology from the University of California at Los Angeles. He resides with his wife, Grace, and their two children in San Francisco.